ARSENIC

DETECTION, MANAGEMENT STRATEGIES AND HEALTH EFFECTS

CHEMISTRY RESEARCH AND APPLICATIONS

Additional books in this series can be found on Nova's website
under the Series tab.

Additional e-books in this series can be found on Nova's website
under the e-book tab.

ARSENIC

DETECTION, MANAGEMENT STRATEGIES AND HEALTH EFFECTS

MARISSA JANE OLSON
EDITOR

New York

For permission to use material from this book please contact us:
Telephone 631-231-7269; Fax 631-231-8175
Web Site: http://www.novapublishers.com

NOTICE TO THE READER

The Publisher has taken reasonable care in the preparation of this book, but makes no expressed or implied warranty of any kind and assumes no responsibility for any errors or omissions. No liability is assumed for incidental or consequential damages in connection with or arising out of information contained in this book. The Publisher shall not be liable for any special, consequential, or exemplary damages resulting, in whole or in part, from the readers' use of, or reliance upon, this material. Any parts of this book based on government reports are so indicated and copyright is claimed for those parts to the extent applicable to compilations of such works.

Independent verification should be sought for any data, advice or recommendations contained in this book. In addition, no responsibility is assumed by the publisher for any injury and/or damage to persons or property arising from any methods, products, instructions, ideas or otherwise contained in this publication.

This publication is designed to provide accurate and authoritative information with regard to the subject matter covered herein. It is sold with the clear understanding that the Publisher is not engaged in rendering legal or any other professional services. If legal or any other expert assistance is required, the services of a competent person should be sought. FROM A DECLARATION OF PARTICIPANTS JOINTLY ADOPTED BY A COMMITTEE OF THE AMERICAN BAR ASSOCIATION AND A COMMITTEE OF PUBLISHERS.

Additional color graphics may be available in the e-book version of this book.

Library of Congress Cataloging-in-Publication Data

Arsenic : detection, management strategies and health effects / editor, Marissa Jane Olson.
   pages cm -- (Chemistry research and applications)
 Includes index.
 ISBN: 978-1-63321-054-7 (softcover)
 1. Arsenic--Environmental aspects. 2. Arsenic--Toxicology. I. Olson, Marissa Jane.
 TD427.A77A743 2014
 571.9'455--dc23
                        2014017890

Published by Nova Science Publishers, Inc. † New York

CONTENTS

PREFACE

There are mainly two types of sources of environmental arsenic; the natural processes and the anthropogenic activities. Weathering of rocks, followed by leaching of metals, leads to the introduction of arsenic into soil and water. Two other important natural processes are the biological activity and the volcanic emissions. Humans are exposed to the toxic arsenic (As) species primarily from food and water. Arsenic is considered one of the most important toxic elements found in the environment. The International Agency for Research on Cancer (IARC) has listed As as a human carcinogen since 1980. The Working Group made the overall evaluation on 'arsenic and inorganic arsenic compounds', based on the combined results of epidemiological studies, carcinogenicity studies in experimental animals and data on the chemical characteristics, metabolism and modes of action of carcinogenicity. As a result of this evaluation, As and inorganic As compounds are considered carcinogenic to humans (Group 1), dimethylarsinic acid and monomethylarsonic acid as possibly carcinogenic to humans (Group 2B) and arsenobetaine and other organic arsenic compounds not metabolized in humans, are not classifiable as to their carcinogenicity to humans (Group 3). This book discusses several topics on arsenic which include arsenic intake and urine as a biomarker of exposure in two endemic arsenic areas in Chile; analytical detection of arsenic species in environmental and biological materials; inorganic arsenic; and the distribution and abundance of arsenic in soils and plants.

Chapter 1 – The aim of this chapter was to study the distribution of total arsenic (t-As) and inorganic arsenic (i-As) in drinking water and foods (raw and cooked) consumed in arsenic-endemic areas. Inorganic arsenic intake from both drinking water and foods was estimated using samples collected from two arsenic endemic areas in the Antofagasta Region of Chile. The t-As contents in drinking water varied in the range 265 ± 13 - 572 ± 21 µg L^{-1} and 0.10 ± 0.10 - 0.25 ± 0.01 µg i-As g^{-1} ww in foods. In both arsenic endemic area considered, majority of the participants interviewed exceeded the FAO/OMS reference intake (149.8 µg i-As/day). In this chapter it used urine as arsenic (As)-biomarker of exposure. Detection of As in these biological samples is indicative of systemic absorption after exposure to it. Several samples of urine collected from subjects interviewed in one of the As-endemic area, showed high levels of t-As in the urine that fluctuated between 78 - 459 ng mL^{-1}. It is known that this situation implies a very high risk for human health and then, it is important to carried out some economic and environmentally strategies for reducing the high levels of As in drinking water and foods. Studies done in arid zones have suggested that collecting plants species on contaminated soils may be effective for selecting potential plants to be used in phytoremediation. The ability to accumulate As from certain plants has generated interest for using them in phytoremediation and/or revegetation for the recovery of contaminated soils. This chapter have showed the potential of four plants collected in the contaminated soils of Antofagasta Region of Chile: *Atriplex atacamensis, Atriplex halimus, Lupinus microcarpus* and *Pluchea absinthioides*. Populations in certain regions, including those in the Second Region in Chile, may present a relatively high intake of i-As, derived first from drinking water and second from high consumption of vegetables and cereals. Excessive exposure to i-As in drinking water and foods can lead elevated urinary-As levels, showing that the toxicological risk to which the inhabitants of small towns of the Antofagasta Region, Chile, have been substantial.

Chapter 2 – Chronic exposure to arsenic primarily causes skin, lung, urinary bladder and liver cancer based on epidemiological studies of many countries. Arsenic is a human multisite carcinogen, but it does not easily induce carcinoma in animal models, which has hindered mechanistic studies of arsenic carcinogenesis in the past. So far, the authors know that arsenic toxicity involves genetic and epigenetic changes, and arsenic can affect reactive oxygen production and oxidative stress, enhanced cell proliferation and modulation of gene expression, reduce DNA repair, increase growth factors, and alter DNA methylation and signal transduction, but the molecular mechanism of arsenic-induced human skin and internal organs carcinogenesis are not completely understood. In the present review, the authors have attempted to evaluate and update the carcinogenesis mechanism of arsenic and its compounds based on the authors' research in malignant transformation models in vitro and the available literature.

Chapter 3 – Groundwater pollution by inorganic arsenicals is common in South Asia, Southeast Asia, and South America. An arsenic-containing drug (Trisenox®) is used for the treatment of acute promyelocytic leukemia. The oxidation state (valence) of an arsenic species changes upon absorption by a living entity, and these species are metabolized by methylation/alkylation. Various chemical forms of arsenic are therefore widely distributed in environmental and biological materials. A simple and rapid method for the routine monitoring and chemical identification of arsenicals, based on their environmental and clinical properties, is needed because the toxicological and physiological behaviors of arsenicals depend not only on their valences but also on their methylation/alkylation states. Numerous methods have been developed for identifying arsenic compounds in environmental and biological materials. In this chapter, analytical methods such as colorimetric and spectrometric determination and chromatographic separation are summarized, and the advantages and disadvantages are pointed out, to enable accurate routine analysis.

Chapter 4 – Arsenic is one of the most toxic trace elements and occurs in both inorganic and organic forms, in many foodstuffs. The inorganic arsenic forms (As(III) and As(V)) were deemed group I carcinogen by the International Agency for Research on Cancer (IARC), and therefore the World Health Organization recommended 15 μg of total inorganic arsenic (t-inAs) per body weight as a provisional tolerable weekly intake. This book chapter deals with the risk assessment of the inorganic arsenic content in different foodstuffs and especially for the more sensitive population groups, such as children. Furthermore, in this chapter there will be presented all the different developed methods, concerning the determination of the inorganic arsenic forms and there will be an overall comparison concerning the ability of the different methods to determine low contents of the toxic arsenic forms in the different food matrices.

Chapter 5 – Context. Recent reports demonstrate that poly (ADP)-ribosylation participates in learning processes, together with other epigenetic mechanisms of chromatin remodcling. The enzyme poly (ADP-ribose) polymerase 1 (PARP-1) has zinc finger domains that are target of inhibition by arsenic, therefore inhibition of this enzyme in the central nervous system may contribute to the cognitive deficits observed in humans after prolonged low level arsenic exposure, especially during development. Objective. The aim of this work was to evaluate *in vitro* and *in vivo* PARP-1 inhibition by arsenite.

Materials and methods. PARP-1 activity was analyzed by quantification of substrate concentration (NAD+) and immunoreactivity for the reaction products (poly (ADP-ribosylated) proteins) in arsenic exposed PC-12 cells (0.1 and 1 µM). The same endpoints were evaluated in the brain of life-long arsenic exposed rats, from gestation through lactation until adult age (3 ppm, drinking water). Results. PARP-1 inhibition was demonstrated through decreased immunoreactivity to PAR polymers and increased concentrations of PARP-1 substrate NAD+ in PC12 cells and in regions of the rat brain involved in learning and memory. Also, exposed animals showed poor performance in an associative memory test. Discussion and conclusion. The demonstrated PARP-1 inhibition in the nervous system of arsenic exposed rats and in PC-12 cells may be related to arsenic- induced cognitive deficits. Besides alterations of DNA methylation, this inhibition is another epigenetic modification associated with exposure to a neurotoxicant.

Chapter 6 – Arsenic (As) has evoked concerns related to the environmental and human health issues for decades. In recent years, the concern has been relocated to the front position as more of the world's population relies on groundwater as a source of clean drinking water, which is reported to be contaminated due to the elevated level of arsenic. Arsenic, being a naturally-occurring element in the Earth's crust, has been commonly found as a trace constituent of rocks, soils, sediments, water, and biota. The natural or anthropogenic or both the activities can elevate arsenic concentrations in groundwater, soils, and sediments to toxic levels and consequently, in the plants. Arsenic exists in multiple oxidation states at earth surface conditions, while arsenite and arsenate are by far the most common As-species found in the environment. The surface processes such as precipitation, dissolution, adsorption, and desorption have been controlled by geochemical parameters, such as pH, Eh, ionic composition, and mineral type, and determine the mobility characteristics of arsenic and the abundance in any given location. The chapter will provide a critical review of the issues related to the distribution and abundance of arsenic in the soils and plants.

In: Arsenic
Editor: Marissa Jane Olson

ISBN: 978-1-63321-054-7
© 2014 Nova Science Publishers, Inc.

Chapter 1

ARSENIC INTAKE AND URINE AS BIOMARKER OF EXPOSURE IN TWO ENDEMIC ARSENIC AREAS IN CHILE: PHYTOREMEDIATION AS A STRATEGY FOR THE RECOVERY OF CONTAMINATED SOILS

Oscar P. Díaz[1], Nelson Núñez[2], Yasna M. Tapia[3], Rafael Arcos[4], Rubén Pastene[5], Dinoraz Vélez[6] and Rosa Montoro[6]*

[1]Departamento de Biología, Facultad de Química y Biología, Universidad de Santiago de Chile, Santiago, Chile
[2]Programa Agrícola CODELCO, Calama, Chile
[3]Facultad de Ciencias Agronómicas, Universidad de Chile, Santiago, Chile
[4]Servicio de Salud de Calama, Calama, Chile
[5]Departamento de Química de los Materiales, Facultad de Química y Biología, Universidad de Santiago de Chile, Santiago, Chile
[6]Instituto de Agroquímica y Tecnología de los Alimentos (IATA-CSIC), Valencia, España

---

* Corresponding author: Email: oscar.diaz@usach.cl.

ABSTRACT

The aim of this chapter was to study the distribution of total arsenic (t-As) and inorganic arsenic (i-As) in drinking water and foods (raw and cooked) consumed in arsenic-endemic areas. Inorganic arsenic intake from both drinking water and foods was estimated using samples collected from two arsenic endemic areas in the Antofagasta Region of Chile. The t-As contents in drinking water varied in the range 265 ± 13 - 572 ± 21 μg L^{-1} and 0.10 ± 0.10 - 0.25 ± 0.01 μg i-As g^{-1} ww in foods. In both arsenic endemic area considered, majority of the participants interviewed exceeded the FAO/OMS reference intake (149.8 μg i-As/day). In this chapter it used urine as arsenic (As)-biomarker of exposure. Detection of As in these biological samples is indicative of systemic absorption after exposure to it. Several samples of urine collected from subjects interviewed in one of the As-endemic area, showed high levels of t-As in the urine that fluctuated between 78 - 459 ng mL^{-1}. It is known that this situation implies a very high risk for human health and then, it is important to carried out some economic and environmentally strategies for reducing the high levels of As in drinking water and foods. Studies done in arid zones have suggested that collecting plants species on contaminated soils may be effective for selecting potential plants to be used in phytoremediation. The ability to accumulate As from certain plants has generated interest for using them in phytoremediation and/or revegetation for the recovery of contaminated soils. This chapter have showed the potential of four plants collected in the contaminated soils of Antofagasta Region of Chile: *Atriplex atacamensis, Atriplex halimus, Lupinus microcarpus* and *Pluchea absinthioides*. Populations in certain regions, including those in the Second Region in Chile, may present a relatively high intake of i-As, derived first from drinking water and second from high consumption of vegetables and cereals. Excessive exposure to i-As in drinking water and foods can lead elevated urinary-As levels, showing that the toxicological risk to which the inhabitants of small towns of the Antofagasta Region, Chile, have been substantial.

INTRODUCTION

Arsenic (As) is a naturally occurring element that exits in the environment a number of different forms, each with its own unique physical, chemical, and toxicological characteristics. Inorganic arsenic (i-As), most often in trivalent form (arsenite, i-As^{III}) or pentavalent form (arsenate, i-As^{V}), is the most abundant form of As in nature, and is commonly present in soil, water, and food (Cohen et al., 2006).

Since arsenic (As) is widely distributed in the environment a large number of populations are exposed to this element around the world. Argentina, Bangladesh, Chile, Mexico and Taiwan are among the countries with the highest levels of As in their drinking water usually the main exposure to this element among humans. (Smith et al., 1998). Food and drinking water are the principal sources of nonocupational exposure to i-As for most populations.

Chronic exposure to i-As through drinking water has been associated with hyperkeratosis, spotted melanosis, skin pigmentation and lung cancer (Pradosh & Anupama, 2002). The toxicity of As in humans is dependent on its chemical form with both arsenite and arsenate being the most toxic, methylarsonate (MA) and dimethylarsinate (DMA) are less toxic, and arsenobetaine (AB), which is mainly found in seafood, being considered non-toxic (Jones, F.T., 2006). In humans, 40% to 70% of As exposure is absorbed, metabolized, and excreted within 48 h (Cohen et al., 2006).

After exposure, biotransformation of As takes place into the body and consists, basically, in a series of successive reductions and methylations. In the past, this biotransformation process was thought to be the As detoxification pathway, but since the methylated metabolites containing As with an oxidation state of +3 are considered to be more genotoxic than their parent inorganic species, the role of this metabolic process is under discussion (Rossman, 2003).

After As intake a certain proportion of the compound is accumulated in different parts of the body. After chronic ingestion of i-As, dermatologic lesions may develop (Hughes, 2006). Approximately 40% and 60% of the ingested form is eliminated through the urine within 1 or 2 days (Vather & Lind, 1986; Johnson & Farmer, 1991). Thus, the measurement of urinary arsenic levels is considered as the most reliable indicator of recent exposure to As. Biomarkers of exposure to As have received great attention. The most common arsenic biomarker of exposure is the analysis of total arsenic (t-As) in urine. Urinary porphyrins have also been proposed as an As biomarker of exposure.

Background levels of urinary As range from 5-50 μg L^{-1} (NRC, 1999). Excessive exposure to i-As in drinking water can lead to urinary levels $\geq$ 700 μg L^{-1} (Brima et al., 2006). If only urinary t-As was determined, exposure to i-As, may be overestimated.

Urinary arsenic does not appear to vary over time, so spot collection of first morning void may be used (Hughes, 2006).

Arsenic is nonessential in plants. In woody plant growing on arsenic polluted soils (74.1-218 mg As/kg). Madejón & Leep (2007) found concentrations of As in leaves between 0.27 and 2.29 mg As/kg. Plants vary in their ability to accumulate and tolerate As, and its mobility is generally low with respect to translocation from roots to aerial part (Zao et al., 2009) except in hyperaccumulator plants. The term hyperaccumulator was first used by Brooks et al. (1977) in relation to plants containing more than 1000 mg kg^{-1} (0.1%) of Ni in dry tissue. It was not until the early 1980´s that it was realized that hyperaccumulators might be used to remediate polluted soils by growing a crop of one of these plants and harvesting it to remove the pollutants (Chaney, 1983). The success of any phytoremediation technique depends upon the identification of suitable plant species that hyperaccumulate heavy metals and produce large amount of biomass using established crop production and management practices.

The phytotoxicity effects commonly observed following As exposure include growth inhibition, chlorophyll degradation, nutrient depletion, and oxidative stress (Marques-García & Cordoba, 2010, Moreno-Jiménez et al., 2008), this later is reflected by an increase in malondialdehyde (MDA), whilst SH groups (thiols) can avoid this oxidative stress by the formation of complex with As(III) (Moreno-Jiménez et al., 2008; Tapia et al., 2013).

The *Atriplex genus* (Chenopodiaceae) denominated saltbush, is one of the most important families of plants in the region of Antofagasta, Chile (Poblete et al., 1991; Saiz et al., 2000). These shrubs xerohalophyte are dominant in many arid and semi-arid regions of the world, particularly in saline, arid soils, and they are used for rehabilitation of degraded lands (Conesa et al., 2007; Manousaki & Kalogerakis, 2009), ornamental plants and revegetating sealed landfills (Ingelmo et al., 1998) and for animal feed (Otal et al., 2010).

Preliminary results in order to identify metal accumulator species, have showed the potential of three terrestrial plant species collected in the contaminated soils of Antofagasta region in Chile to extract some of the pollutants especially As. *Atriplex atacamensis, Lupinus microcarpus* and *Pluchea absinthioides* are important native plants that constitute part of the desert scrub community of the pre-Andean area of the Antofagasta Region (Díaz et al., 2011).

MATERIALS AND METHODS

Study Area

Study area include the village of Chiu Chiu (22° 20´ S; 68° 39´ W), Lasana (22° 16´ S; 68° 38´ W), Ayquina (22° 16' S; 68° 19' W) and Socaire (23° 35´ S; 67° 53´ W), located at 35, 45, 74 and 180 km to the northeast of Calama city, respectively (Figure 1), which were classified as arid soils.

The zone of pre-Andean Antofagasta region is influenced by the Loa river and is located prior to the confluence with the Salado river (Figure 1). The zone is characterized by scarce/poor plant cover, saline properties of soils and high concentrations of As of natural origin.

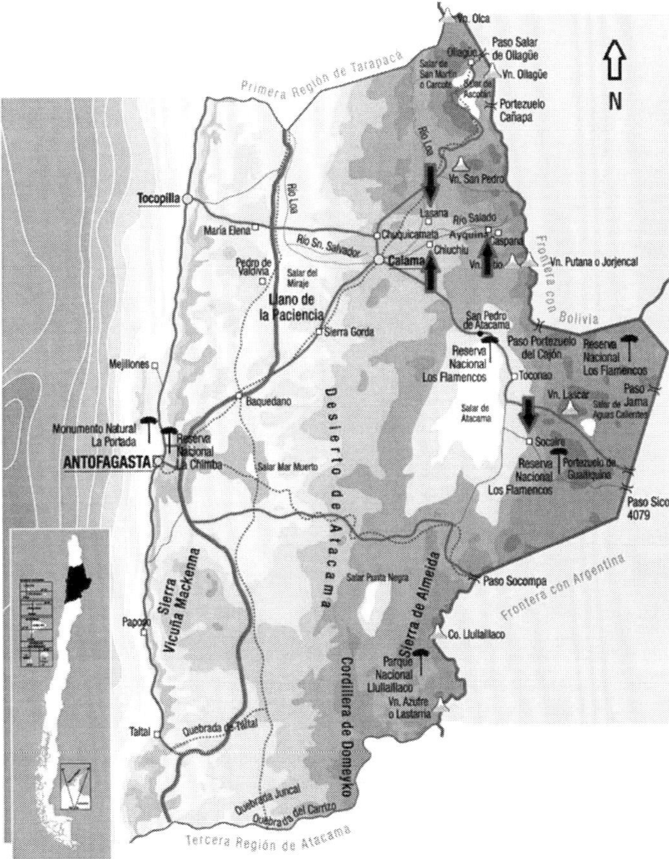

Figure 1. Study area of Chiu Chiu, Lasana, Ayquina, and Socaire villages.

Food Questionnaire

A 24-hour dietary recall questionnaire was given to residents in the township of Chiu Chiu and Socaire. The questionnaire asked for information about the type and quantity of water consumed, foods ingested and how the food were prepared for consumption, raw or cooked, demographic characteristic such as age, sex and information on potencial exposures sources, suh as source of drinking water, smoking patterns, location of residence, and occupation. The design of the questionnaire and the method for performing the dietary study was approved by the Regional Ministerial Health Secretariat for Chile´s Antofagasta Region. Professionals from Health Service in Calama and San Pedro de Atacama (Antofagasta Region), and from the Chile Second Region Government CODELCO Agreement Agricultural Program, besides our research group administered the questionnaire.

The number of interviewees from Chiu Chiu was set at 50, 25 male and 25 female belonging to different age groups. A total of 21 participants adults, 7 male and 14 female were selected from the township of Socaire. The mean body weight was 70 Kg and the mean age of 40.2 years. All of the participants had been living in the villages for over 5 years.

Water and Food Samples

The drinking water and food samples (raw and cooked), were collected in two sampling period during 2000 in Chiu Chiu and 2008 and 2009 from Socaire. Both samples were obtained directly from the homes where the interviewees lived. Drinking water samples were collected in acid-washed polyethylene bottles from the kitchen tap (or other designated drinking water source, excluding bottled water) after allowing the water to run through the pipes for several seconds. The bottle was filled to the top and the label completed and affixed. The water samples were placed in a refrigerated container for transport. Control samples were obtained from 5 homes located in Calama, one of the main cities in the Antofagasta Region that now has a water treatment plant. The water from this city contains the maximum permitted by the FAO/OMS (10 μg As L^{-1}). To preserve the As in the water samples, 0.5 mL of HCl (0.01 mol L^{-1}) was added to each sample. The water samples were transferred to the laboratory and were stored at 4 °C until the As analysis.

Most food products were grown in the communities of Chiu Chiu and Socaire. Once foods were collected (raw and cooked) a sample mixture of these foods was placed in an aluminium box of 500 g capacity, frozen at 20 °C and then freeze-dried. The lyophilized samples were ground in a domestic apparatus, and the resulting powder was vacuum packed and stored at 4 °C until the As analysis.

Urine Sample Collection and Preparation

Participants from Socaire were supplied a polyethylene bottle and instructions for urine collection prior to the day of the interview. Participants were asked to collect a first morning urine sample directly into polyethylene bottles and whether to adjust the data to volume of urine voided (related to urinary creatinine levels or specific gravity). Concentrated HCl (1 mL to 100 mL urine) was added to prevent bacterial growth. The samples were then stored frozen at -20 °C until the analyses were carried out, except during the transportation when they were kept on wet ice. Prior to analysis, the samples were filtered through a 0.45 µm syringe filter and diluted up to 5-fold with 2% HNO_3 for t-As determination.

Zone of Soil and Terrestrial Plants Collection and Determination of the Main Physico-Chemical Properties

Native plants and soil samples were collected from three pre-Andean zones in the Region of Antofagasta (Chile) between 2004 until 2012. These zones were (1) Chiu Chiu (2) Lasana and (3) Ayquina (Figure 1). The climate of the study zones is characterized by extreme aridity throughout the year (rainfall 5.7 mm per year) with very low relative humidity (average 33%). The average temperature is 20 °C in summer and winter. Due to these unfavorable conditions, the presence of vegetation is scarce and therefore the number of individuals collected was limited.

Surface soil samples (0-20 cm depth) were collected from the area adjacent to the plants roots. The pH was determined in aqueous extract 1:5. Percentage slit clay was determined by weighing the soil retained on a sieve with a diameter > 0.080 mm. Organic matter was determined by a colorimetric method using $K_2Cr_2O_7$, H_2SO_4 and saccharose solution as standard solutions. For a determination of total mineral elements, soil samples were dried at 40°C to constant weight, sieved to particle size of 2 mm and digested with aqua regia (HNO_3-HCl,1:3). Total Ca, K, Mg, N, P, As, B, Cu, Fe, Mn, and Zn were determined by atomic absorption spectrometry (AAS). To determine total N, the Kjeldahl method was used. Total P and S were determined by elemental analysis. Boron concentration was determined using a colorimetric method with azomethine after extraction with hot water. For determination total As concentrations in the soil, dry ashing mineralization and quantification by flow injection hydride generation atomic absorption spectrometry (FI-HG-AAS) were employed. Bioavailable As concentration was measured by FI-HG-AAS after extraction with $NaHCO_3$. Detection limit of the apparatus is 0.006 $\mu g\ g^{-1}$ dw. Accuracy for Montana Soil (certified standard) found value = 98.19 ± 0.03 $\mu g\ g^{-1}$ (certified value = 105 ± 8 $\mu g\ g^{-1}$).

Four native plant species were collected in triplicate from their natural habitat: *Pluchea absinthioides* was collected from Chiu Chiu, *Atriplex atacamensi* and *Atriplex halimus* were collected from Lasana and Chiu Chiu respectively and *Lupinus microcarpus* was collected from Ayquina. Plants were separated into leaves, stems and roots, dried at 60 ± 5 °C until constant weight and then ground and sieved. Later samples were calcined and t- As concentration was determined by FI-HG-AAS. All results are expressed in dry weight (dw). The bioconcentration factor (BCF) and transport index (TI) of As to the leaf were calculated.

The data were anayzed using single-factor ANOVA in the statistical program SPSS versión 13.0. To test differences in bioavailable As concentrations between soil samples and t-As concentrations between leaves, stems and roots, Duncan´s test was used with a significative level of $p \leq 0.05$.

RESULTS AND DISCUSSION

The concentration of total and inorganic As found in the drinking water and foods are shown in Table 1.

The t-As concentrations in the drinking water obtained in Chiu Chiu village showed an average value of 572 ± 21 $\mu g\ L^{-1}$ and 40 ± 0.2 $\mu g\ L^{-1}$ in the first and the second sampling period, respectively, and 265 ± 13 $\mu g\ L^{-1}$ and 293 ± 44 $\mu g\ L^{-1}$ in the first and the second sampling period, respectively, carried out in Socaire. All of these concentration values show a low variation in the results and were much higher than those obtained in the control samples from Calama.

The first sampling of drinking water were collected during 2000´s, and the mean i-As content in the water collected from various homes in the village of Chiu Chiu was 572 µg L^{-1} (Table 1). This concentration was over 10 times greater than the level permitted by Chilean legislation in that moment (50 µg L^{-1}). As a result of the preliminary results obtained in that work, the administrative authority of the Antofagasta Region in Chile decided to use water tankers to provide Chiu Chiu with drinking water that conformed to Chilean regulations. Consequently, in the samples collected in the second period, the i-As content quantified in the water used by the population for drinking and cooking was much lower (40 µg L^{-1}).

Table 1. Total and inorganic arsenic concentration in the drinking water (mean, µg L^{-1}) and food (mean, µg g^{-1}, ww) from the first and second sampling period

Period	Drinking Water (µg As L^{-1})		Food (µg g^{-1}, ww)			
			t-As		i-As	
	Chiu Chiu	Socaire	Chiu Chiu	Socaire	Chiu Chiu	Socaire
1st Sampling ($\bar{x}\pm$SD)	572±21	265±13	0.28±0.47	0.29±0.02	0.24±0.42	0.25±0.01
2^{n} Sampling ($\bar{x}\pm$SD)	40±0.2	293±44	0.10±0.11	0.47±0.20	0.10±0.10	0.23±0.09
Control	40±0.1	6.7±0.3				

These values were also much higher than the máximum limit of arsenic levels in drinking water established by the FAO/OMS (10 µg L^{-1}), demostrating a high level of pollution in the drinking water of Chiu Chiu and Socaire. The differences between Calama and Chiu Chiu and Socaire are because the drinking water in Calama is treated in an As-abatement plant.

In the Antofagasta Region (Chile), most of the small towns do not have water treatment plants, and consequently the population consumes water with As contents greater than the maximum permitted by Chilean legislation (actually 10 µg L^{-1}). These results agreed with those of Bae et al. (2002) who measured As concentrations between 223 to 372 µg L^{-1} in the drinking water.

The concentrations of t-As and i-As found in the foods (raw and cooked) are shown in Table 1. The t-As concentrations varied between 0.28 ± 0.47 and 0.10 ± 0.11 µg g^{-1} ww in the foods collected in Chiu Chiu, during the first and the second sampling period, respectively. The concentration of i-As varied between 0.24 ± 0.42 and 0.10 ± 0.10 µg g^{-1}, ww in this place and both period, respectively.

The t-As concentrations varied between 0.29 ± 0.02 and 0.47 ± 0.20 µg g^{-1} in the foods collected in Socaire, during de first and the second sampling period, respectively. The concentration of i-As varied between 0.25 ± 0.01 and 0.23 ± 0.09 µg g^{-1} ww in this place and both period, respectively.

During the first sampling period are not differences between t-As and i-As concentrations in the foods collected both Chiu Chiu and Socaire. However during the second sampling period it is observed differences between t-As and i-As levels in the foods obtained both Chiu Chiu and Socaire. In the first sampling period, the water used for drinking and cooking purposes in Chiu Chiu, contained 572 µg As L^{-1}, while in the second sampling period, the water contained 40 µg As L^{-1} (Table 1).

Few studies have assessed the effects of the cooking process on t-As and As speciation in Latin America. Health risk evaluations of As intake are generally base don the As contents of raw products. However, food is often consumed after being processed and cooked, which could alter the t-As content as well as its chemical speciation (Bundchuh et al., 2012).

The variation in the t-As and i-As concentrations observed in the food could be explained by the type of food tested and the specific preparation procedure used, in which the use of water is a very important factor because of the high levels of As in the water. The locally produced foods, such as vegetables and quinoa, were a regular part of the diet as determined by the food questionnaire. These food samples showed higher As levels than those foods samples that were not produced locally, such as noodles, oil and beef. The As levels increased if these foods were prepared using certain cooking process. Similar findings were also described by Queirolo et al. (2000) and Laparra et al. (2005).

From the data provided by the 24 h recall questionnaire administered to the inhabitants of Chiu Chiu and Socaire (quantity of water, L/day and food ingested, grams ww) and the t-As and i-As concentrations found in the drinking water, raw and cooked food analyzed it was calculated the intakes of i-As in each period of sampling. The determination of t-As has no toxicological interest, but its inclusion in this chapter makes it possible to compare the contamination of the foods analyzed with the data available in the literature, which almost all analyze t-As and not i-As. These intakes, expressed as µg/day, are shown in Table 2. To evaluate the health risks of this estimated dietary exposure, the values were compared to the Provisional Tolerable Weekly Intake (PTWI) values of 15 µg i-As/week/kg body weight recommended by the Joint FAO/WHO Expert Committee for Food Additives (JECFA) (Del Razo et al., 2002; Bundschuh et al., 2012).

To compare the PTWI with the data obtained for the inhabitants of Chiu Chiu and Socaire, the procedure carried out by Díaz et al. (2004) and Anawar et al. (2012), we used in this chapter and the following two approximations. First, as the questionnaire administered was a 24 h dietary recall, the PTWI is expressed as a tolerable daily intake (TDI=PTWI/7 days) with a value of 2.14 µg i-As/kg body weight. Second, all of interviewees were adults that informed a mean body weight of 70 Kg. Assuming this body weight, the reference intakes stated by the FAO/OMS are equivalent to 149.8 µg i-As/day (TDI x Kg body weight). This chapter compared this value to the i-As intake estimated for the inhabitants of Chiu Chiu and Socaire (Díaz et al., 2004).

Table 2. Dietary inorganic arsenic intake from water (µg /day) and food (µg /day, ww) consumed by adult interviewees in Chiu Chiu and Socaire villages, Antofagasta Region, Chile

Statistics	First Sampling Period (µg i-As /day)		Second Sampling Period (µg i-As /day)	
	Chiu Chiu	Socaire	Chiu Chiu	Socaire
Mean	1378	660	125	564
Range	475-1647	181-1370	57-200	360-862

The i-As intake for all of the interviewed individuals from the two sampling periods, showed a great range of variability, especially during the first sampling period, this was partially due to the variable consumption of water. The variability was also influenced by the volumes and types of foods, in particular the vegetables and cereals in which i-As is the major species. Previous studies in other arsenic-endemic areas, in which the food groups considered in the intake estimate were similar to those analysed in Chile, estimated the i-As intake by an extrapolation from the t-As intake, assuming that i-As represented all of the t-As measured (Roychowdhury et al., 2002) or at least 50% of the t-As measured (EPA, 1997).

The FAO/OMS reference intake was exceeded by all of those interviewees, excepted interviewees from Chiu Chiu, during the second sampling period.

The above mentioned results are agree with the total arsenic concentration found in urine collected from adult inhabitants of Socaire village, which showed high levels of t-As in this biomarker both first sampling period (Mean:

259 ng mL^{-1}) as in the second sampling period (mean: 209 ng mL^{-1}) (Table 3). These values were much higher than those obtained in the control samples from Calama (mean: 28 ng mL^{-1}).

The differences between Calama and Socaire are because the drinking water in Calama contains As concentration lower than water consumed by the inhabitants of Socaire. It is known that the drinking water usually is the main exposure to this element among humans (Morgan, 2001). Approximately 40% and 60% of the ingested form is eliminated through the urine within 1 or 2 days. Thus the measurement of urinary arsenic levels is considered as the most reliable indicator of recent exposure to As (Johnson & Farmer, 1991).

Background levels of urinary As range from 5 to 50 ng mL^{-1} (NRC, 1999). The ACGIH (2000) Biological Exposure Indices (BEI) recommended value for urinary arsenic level has been revised to 35 ng mL^{-1}, as modified from the previous value of 50 µg g creatinine for arsenic established in 1997 (Yaw-Huei et al., 2002).

Table 3. Total arsenic concentration in urine (ng /mL^{-1}) collected from adult interviewees in Socaire village, Antofagasta Region, Chile

Statistics	First Sampling Period (ng t-As mL^{-1})	Second Sampling Period (ng t-As mL^{-1})	Control (ng t-As mL^{-1})
Mean	259 (n=17)	209 (n=9)	28 (n=4)
Range	78-459	84-416	11-38

The results presented in this chapter, indicate that the toxicological risk to which the inhabitants of Socaire have been exposed has been substantial for many years and then it is important to carry out some economic and environmentally strategies for reducing the high levels of As ingested. Studies done in arid zones have suggested that certain plant species on contaminated soils may be effective for reducing As levels both agricultural soils and vegetable foods.

Physico-chemical characteristics and total and bioavailable arsenic concentrations in soils collected from three arid zones near of the city of Calama, are presented on Table 4.

The pH for Chiu Chiu and Lasana soil was similar (pH 8.5 and pH 8.3, respectively), while Ayquina soil had the lowest pH (pH 7.6). Ayquina soil contained a higher percentage of silt-clay (4.33%) than soils from Chiu Chiu (2.66%) and Lasana (3.66%). In general, all of the soils were dominated by a sandy texture. Ayquina soil had a higher concentration of organic matter and of Ca, N, P, and Fe than did Chiu Chiu and Lasana soils. *Lupinus microcarpus* a nitrogen-fixing legume, which could explain the higher concentration of N in this soil. Levels of B, Cu, Mn, and Zn are low in the three soils.

Table 4. Main physico-chemical soil characteristic and total and bioavailable arsenic concentration (mg kg^{-1} dry matter) in soil

	Control Soil	Zones		
		Chiu Chiu	Lasana	Ayquina
pH	7.8 ± 0.1	8.5 ± 0.3	8.3 ± 0.3	7.6 ± 0.3
Slit-clay (%)	25 ± 21	2.66 ± 1.5	3.66 ± 1.5	4.33 ± 1.1
Ca (%)	nd	0.50 ± 0.34	0.87 ± 0.36	1.40 ± 0.14
K (%)	nd	0.06 ± 0.02	1.04 ± 0.01	1.12 ± 0.16
Mg (%)	nd	0.78 ± 0.33	1.55 ± 0.28	0.98 ± 0.03
N (%)	0.27 ± 0.02	0.03 ± 0.02	0.02 ± 0.01	0.13 ± 0.07
P (%)	0.21 ± 0.0	0.04 ± 0.01	0.04 ± 0.02	0.12 ± 0.02
As (mg kg^{-1})	12.7 ± 1.1	52.98 ± 9.91	54.34 ± 15.43	5.58 ± 1.04
Bioavailable As (mg kg^{-1})	nd	0.22 ± 0.09	0.15 ± 0.05	0.01 ± 0.005
B (mg kg^{-1})	4.13 ± 0.06	0.23 ± 0.15	0.11 ± 0.01	0.02 ± 0.01
Cu (mg kg^{-1})	431 ± 21	1.63 ± 1.18	0.27 ± 0.06	1.03 ± 0.32
Fe (mg kg^{-1})	30 ± 25	13.63 ± 9.49	9.73 ± 5.22	16.66 ± 2.89
Mn (mg kg^{-1})	85.8 ± 10.4	4.54 ± 0.60	4.23 ± 1.00	3.50 ± 1.04
Zn (mg kg^{-1})	1700 ± 5.2	1.00 ± 0.44	1.06 ± 0.64	0.33 ± 0.21

Values are means ± standard desviations, n = 4.

Total As concentration in soils ranged from 41.12 to 65.72 mg kg^{-1} (mean: 52.98 mg kg^{-1}) for Chiu Chiu, from 32.59 to 68.23 mg kg^{-1} (mean: 54.34 mg kg^{-1}) for Lasana and from 4.24 to 7.28 mg kg^{-1} (mean: 5.58 mg kg^{-1}) for Ayquina. These results indicate that Chiu Chiu and Lasana soils are contaminated with As, whereas Ayquina soils, with levels of As lower than 6 mg kg^{-1} are consistent with values for uncontaminated soils.

Bioavailable As concentrations in Chiu Chiu and Lasana soils were 0.22 and 0.15 mg kg^{-1}, respectively, which represented 0.42% and 0.27% of the total As concentrations, respectively. In Ayquina, the soil bioavailable As concentration was 0.01 mg kg^{-1}, which represented 0.18% of the total As concentration (Table 1). Chemical reactivity of As in solution is primarily controlled by surface reactions, specifically complexation with oxides/hydroxides of Al, Mn, and especially Fe (Fitz & Wenzel, 2002). Fe oxides are the primary adsorption sites for As in most soils and sediments, thereby decreasing its availability (Visoottiviseth et al., 2002). pH also is one of the main factors determining the bioavailability of As in contaminated soils (Fayiga et al., 2007; Díaz et al., 2011). In general, high pH values increased bioavailable As in soil solution (Vásquez et al., 2008). In some cases, arsenic soil retention was observed to increase when organic matter and clay content were higher (Balasoiu et al., 2001). The lowest bioavailable As concentration found in Ayquina soils may be associated with the low t-As concentration and higher organic matter and clay content than in Chiu Chiu and Lasana soils. The results indicate that t-As concentrations in arid soils of Lasana and Chiu Chiu, where *A. atacamensis, A. halimus* and *P. absinthioides* were collected respectively, reached pollutant levels. By contrast, in Ayquina soils, where *L. micocarpus* was collected, As levels did not indicate pollution.

Physico-chemical values founded in control soil are higher much than values obtained in soils recolected from study areas, except As levels measured in Ayquina soil were lower than control soil. This result show that the physico-chemical characteristics in soils vary regarding the study area (Díaz et al., 2011; Tapia et al., 2013).

Table 5 shows the dry weight and As concentration in leaves, stems and roots of *P. absinthioides, A.halimus, A. atacamensis* and *L. microcarpus* collected in different zones. Dry weight was similar between the species. Arsenic concentration was higher in leaves than in roots, except in *A. halimus*. These results are uncommon because most plants preferentially accumulate As in their roots, as in the case of *A. halimus* and, even if they absorb high levels, transport of the metalloid to the aerial parts is minimal (Jedynak et al., 2009).

In general, As levels found in leaves of all plant species (5.5-9.7 mg kg^{-1}) was high compared to other species. For example, in the herbaceus species *Cameroon angustifolium, Cirsium ssp* and *Reseda luteola*, which were collected from soils (pH 7.3) with As levels similar to those found in Chiu Chiu and Lasana soils, As concentrations in leaves were between 1.01 and 2.90 mg kg^{-1} (Madejón and Leep, 2007).

The highest As concentration in leaves was found in *L. microcarpus* (9.7 ± 1.6 mg kg^{-1}). However, As concentrations in Ayquina soils where this plant species was collected, were lowest. Other researchers have also found high levels in plants collected from soils with low As levels (Visoottiviseth et al., 2002).

Bioconcentration factor (BCF) was considerably higher in *L. microcarpus* (BCF: 1.8, Table 5), indicating that this specie has an important ability to accumulate As. Some researchers have suggested that phytoextraction is feasible when the BCF is > 1 (Zao et al., 2009).

Transpor index (Ti) of As to leaves was higher than in all species, except *A. halimus*, therefore plants accumulated As in leaves. *L. microcarpus* reached the highest Ti, showing a remarkable ability to translocate As to the leaf. It would be important in future to carried out studies to evaluate the response of *L. microcarpus* to soils contaminated with arsenic and to investigate its possible use in phytoextraction.

CONCLUSION

Inhabitants of Chiu Chiu and Socaire consume drinking water and food with high As levels primarily i-As, depending of the type of food. The high As content in the food is primarily derived from the drinking water, which contains high As concentrations. Majority of the interviewed participants exceeded the FAO/OMS reference intake (149.8 µg i-As/day). High t-As levels in urine collected from adult inhabitants of Socaire village were obtained (Mean: > 200 ng mL^{-1}). This value was much higher than those obtained in the control samples and the recommended value for urinary As level (35 ng mL^{-1}), showing that the toxicological risk to which the inhabitants of Socaire have been exposed has been substantial.

Tabla 5. Dry weight (g) and As concentration (mg kg^{-1} dry matter), bioconcentration factor (BCF) and transport index (Ti) of As to the leaves

Native plant species	Plant organ	Dry weight (g)	As (mg kg^{-1})	BCF	Ti
P. *absinthioides* (Chiu Chiu zone)	Leaf	2.7 ± 0.2	5.9 ± 1.0	0.1	1.7
	Stem	2.8 ± 0.9	5.1 ± 0.4		
	Root	2.6 ± 0.7	3.4 ± 0.7		
A.*Halimus* (Chiu Chiu zone)	Leaf	nd	5.45 ± 1.76	0.05	0.80
	Stem	nd	1.50 ± 0.20		
	Root	nd	6.54 ± 0.55		
A. *atacamensis* (Lasana zone)	Leaf	2.1 ± 0.7	6.3 ± 0.7	0.1	3.7
	Stem	2.5 ± 0.7	2.5 ± 0.7		
	Root	2.3 ± 0.3	1.7 ± 0.4		
L. *microcarpus* (Ayquina zone)	Leaf	2.3 ± 0.3	9.7 ± 1.6	1.8	6.1
	Stem	2.4 ± 0.5	2.5 ± 0.9		
	Root	1.5 ± 0.2	1.6 ± 0.4		

Finally, the four native plant species resist contamination by As and can be recommended its possible use in phytoremediation soils. Besides, to generate plant cover in As contaminated soils under saline conditions controlled, preventing the dispersion of this metalloid and others toxic elements via wind and leaching.

ACKNOWLEDGMENTS

This chapter was a collaboration between the Universidad de Santiago de Chile (USACH-DICYT), Ministerio del Interior (Gobierno de Chile), Programa Agrícola de la Corporación del Cobre (CODELCO), Chile, Programa Iberoamericano de Ciencia y Tecnología para el Desarrollo (CYTED), España, Instituto de Agroquímica y Tecnología de Alimentos, Consejo Superior de Investigación Científica (IATA-CSIC), España, CSIC-USACH Project: Estudio de la idoneidad del uso de metabolitos urinarios de arsénico inorgánico como herramienta para la evaluación del riesgo en poblaciones expuestas, 2008-2009. Professor Glauco Morales from University of Antofagasta (Chile), Psycologists Pedro Alvarado and Jessica Parada (Chile), and Dra. Pilar Bernal (CEBAS-CSIC), España.

REFERENCES

ACGH. Treshold limit values for chemical substances and physical agents and biological exposures indices. *American Conference of Governmental Industrial Hygienists*, Cincinnati, OH. 2000.

Anawar, H., García-Sánchez, A., Hossain, N., Akter, S. Evaluation of health risk and arsenic levels in vegetables sold in markets of Dhaka (Bangladesh) and Salamanca (Spain) by AAS-GH. *Bull. Environ. Contam. Toxicol.*, 2012, 89, 620-625.

Bae, M; Watanabe, C; Inaoka, T; Sekiyama, M; Sudo, N; Bokul, MH; Ohtsuka, R. Arsenic in cooked rice in Bangladesh. *Lancet.*, 2002, 360, 1839-1840.

Balasoiu, C; Zagury, G; Deschenes, L. Partitioning and speciation of chromium, copper, and arsenic in CCA-contaminated soils: Influence of soil composition. *Sci. Total Environ.*, 2001, 280, 239-255.

Brima, EI; Haris, PI; Jenkins, RO; Polya, DA; Gault, AG; Harrington, ChF. Understanding arsenic metabolism through a comparative study of arsenic levels in the urine, hair and fingernails of healthy volunteers from three unexposed ethnic groups in the United Kingdom. *Toxicology and Applied Pharmacology*, 2006, 216, 122-130.

Brooks, RR; Lee, J; Reeves, RD; Jaffré, T. Detection of nickeliferous rocks by analysis of herbarium specimens of indicator plants. *J. Geochem. Explor.*, 1977, 7, 49-57.

Bundschuh, J; Nath, B; Battacharya, P; Lui, Ch; Armienta, MA. Arsenic in the human food chain: The Latin America perspective. *Sci. Total Environ.*, 2012, 429, 92-106.

Chaney, RL. Plant uptake of inorganic waste constitutes. IN: Parr, JK; Marsh, PB; Kla, JM (Eds.). *Land Treatment of Hazardous Wastes*. Noyes Data, Park Ridge, IL. 1983, 50-76.

Cohen, SM; Arnold, LL; Eldan, M; Lewis, AS; Beck, BD. Methylated Arsenicals: The implications of metabolism and carcinogenicity studies in rodents to human risk assessment. *Critical Reviews in Toxicology*, 2006, 133, 36-99.

Conesa, H; García, G; Faz, A; Arnaldos, R. Dinamics of metal tolerant plant communities development in mine tailings form the Cartagena-La union Mining District (SE, Spain) and their interest for further revegetation purposes. *Chemosphere*, 2007, 68, 1180-1185.

Del Razo, LM; García- Vargas, GG; García- Salgado, J; Sanmiguel, MF; Rivera, M; Hernández, MC; Cebrián, ME. Arsenic levels in cooked food and assessment of adult dietary intake of arsenic in the Region Lagunera, Mexico. *Food Chem. Toxicol.*, 2002, 40, 1423-1431.

Díaz, O; Tapia, Y; Pastene, R; Montes, S; Núñez, N; Vélez, D; Montoro, R. Total and bioavailable arsenic concentration in arid soils and its uptake by native plants from pre-Andean zones in Chile. *Bull. Environ. Contam. Toxicol.*, 2011, 86, 666-669.

Díaz, OP; Leyton, I; Muñoz, O; Núñez, N; Devesa, V; Suñer, MA; Vélez, D; Montoro, R. Contribution of water, bread and vegetables (raw and cooked) to dietary intake of inorganic arsenic in a rural village of Northern Chile. *J. Agric. Food Chem.*, 2004, 52, 1773-1779.

Fayiga, AO; Ma, LQ; Zhou, Q. Effects of plants arsenic uptake and heavy metals on arsenic distribution in an arsenic-contaminated soil. *Environ. Pollut.*, 2007, 147, 737-742.

Fitz, WJ; Wenzel, WW. Arsenic transformations in the soil- rhizosphere-plant system: Fundamentals and potential application to phytoremediation. *J. Biotechnol.*, 2002, 99, 259-278.

Hughes, MF. Biomarkers of exposure: A case study with inorganic arsenic. *Environ. Health Perspect.*, 2006, 114, 1790-1795.

Hwang, Y-H; Lee, Z-Y; Wang, J-D; Hsueh, Y-M; Lu, I-Ch; Yao, W-L. Monitoring of arsenic exposure with speciated urinary inorganic arsenic metabolites for ion implanter maintenance engineers. *Environmental Research*, 2002, 90, 207-216.

Ingelmo, F; Canet, R; Ibáñez, MA; Pomares, F; García, J. Use of MSW compost dreed sewage sludge and other wastes as a partial substitutes for peat and soil. *Bioresour. Technol.*, 1998, 63, 123-129.

Jedynak, L; Kowalska, J; Harasimowicz, J; Golomowski, J. Speciation analysis of arsenic in terrestrial plants from arsenic contaminated area. *Sci. Total Environ.*, 2009, 407, 945-952.

Johnson, LR; Farmer, JG. Use of human metabolic studies and urinary arsenic speciation in assessing arsenic exposure. *Bull Environ Contam Toxicol*, 1991, 46, 53-61.

Jones, FT. A broad view of arsenic. *Poultry Science*, 2006, 86, 2-14.

Laparra, JM; Vélez, D; Barbera, R; Farré, R; Montoro, R. Bioavailability of inorganic arsenic in cooked rice: Practical aspects for human risk assessments. *J. Agric. Food Chem.*, 2005, 53, 8829-8833.

Madejón, P; Leep, N. Arsenic in soils and plants of woodland regenerated on an arsenic-contaminated substrate: A sustainable natural remediation. *Sci. Total Environ.*, 2007, 379, 256-262.

Manousaki, E; Kalogerakis, N. Phytoextraction of Pb and Cd by the Mediterranean salt bush (*Atriplex halimus*): Metal uptake in relation to salinity. *Environ. Sci. Pollut. Res.*, 2009, 16, 844-854.

Marques-García, B; Córdoba, F. Antioxidative system in wild populations of *Erica andevalensis*. *Environ Exp Bot.*, 2010, 68, 58-65.

Moreno-Jiménez, E; Peñaloza, J; Carpena, R; Esteban, E. Comparison of arsenic resistance in Mediterranean woody shrubs used in restoration activities. *Chemosphere*, 2008, 71, 466-473.

Morgan, A. Exposure and Health Effects. Office of Water, Science and Technology, Health and Ecological Criteria Division, U.S. Environmental Protection Agency, Washington, USA, (Chapter 3). 2001.

National Resarch Council (NRC). Subcommittee on Arsenic in Drinking Water. Arsenic in Drinking Water. National Academy Press. Washington D.C. 1999.

Otal, J; Orengo, J; Quiles, A; Hevia, M; Fuentes, F. Characterization of edible biomass of *A. halimus* and its effects on feed and water intakes and on blood mineral profile in non- pregnant Manchega-breed sheeps. *Small Rumin. Res.*, 2010, 91, 208-214.

Poblete, V; Campos, V; González, L; Montenegro, G. Anatomical leaf adaptations in vascular plants of salt marsh in the Atacama Desert (Chile). *Rev. Hist. Nat.*, 1991, 64, 65-75.

Pradosh, R; Anupama, S. Metabolism and toxicity of arsenic: A human carcinogen. *Current Science*, 2002, 82 (1), 38-45.

Queirolo, F; Stegen, S; Mondaca, J; Cortés, R; Rojas, R; Contreras, C; Muñoz, L; Schwuger, MJ; Ostapczuk, P. Total arsenic, lead, cadmium, copper and zinc in some salt rivers in the Northern Andes of Antofagasta, Chile. *Sci. Total Environ.*, 2000, 255, 85-95.

Rossman, TG. Mechanism of arsenic carcinogenesis: An integrated approach. *Mutat. Res.*, 2003, 533, 37-65.

Roychowdhury, T; Uchino, T; Tokunaga, H; Ando, M. Survey of arsenic in food composites from an arsenic- affected area of West Bengal, India. *Food Chem. Toxicol.*, 2002, 40, 1611-1621.

Saiz, F; Yates, L; Núñez, C; Daza, M; Varas, M; Vivar, C. Biodiversidad de complejo de artrópodos asociados al follaje de la vegetación del norte de Chile, Segunda Región. *Rev. Chil. Hist. Nat.*, 2000, 73, 671-692.

Smith, AA; Goycolea, M; Haque, R; Biggs, ML. Marked increase in bladder and lung cancer mortality in a region of Northern Chile due to arsenic in drinking water. *Am. J. Epidemiol.*, 1998, 147, 660-669.

Tapia, Y; Díaz, O; Pizarro, C; Segura, R; Vines, M; Zúñiga, G; Moreno-Jiménez, E. *Atriplex atacamensis* and *Atriplex halimus* resist As contamination in Pre-Andean soils (northern Chile). *Sci. Total Environ.*, 2013, 450-451, 188-196.

U.S. Environmental Protection Agency (EPA). Health Effects Assessment. Summary Tables: FY update (HEAST). EPA/540/2-97/036. 1997.

Vásquez, S; Carpena, R; Bernal, MP. Contribution of heavy metals and As-loaded lupin root mineralization to the availability of the pollutants in multi-contaminated soils, 2008, *Environ. Pollut.*, 152, 373-379.

Vather, M; Lind, B. Concentrations of arsenic in urine of the general population in Sweden. *Sci. Total Environ.*, 1986, 54, 1-12.

Visoottiviseth, P; Francesconi, K; Sridokchan, W. The potential of Thai indigineous plant species for the phytoremediation of arsenic contaminated land. *Environ. Pollut.*, 2002, 118, 453-461.

Hwang, H.Y., Lee, Z.Y., Wang, J.D., Hsueh, Y.M., Ku, I.C., Yao, W.L. Monitoring of arsenic exposure with speciated urinary inorganic arsenic metabolites for ion implanter maintenance engineers. *Environmental Research*, 2002, 90, 207-216.

Zao, F; Ma, F; Meharg, AA; Mc Grath, SP. Arsenic uptake and metabolism in plants. *New Phytol.*, 2009, 181, 777-794.

In: Arsenic

Editor: Marissa Jane Olson

ISBN: 978-1-63321-054-7

© 2014 Nova Science Publishers, Inc.

Chapter 2

MALIGNANT TRANSFORMATION MODELS IN VITRO FOR THE STUDY OF ARSENIC CARCINOGENESIS

Ning Ma[1] and Shiwen Huang[2]*

[1]Faculty of Health Science, Suzuka University of Medical Science,
Suzuka, Mie, Japan

[2]Guangxi Zhuang Autonomous Region Institute for the Prevention and
Treatment of Occupational Disease, Nanning, Guangxi, China

ABSTRACT

Chronic exposure to arsenic primarily causes skin, lung, urinary bladder and liver cancer based on epidemiological studies of many countries. Arsenic is a human multisite carcinogen, but it does not easily induce carcinoma in animal models, which has hindered mechanistic studies of arsenic carcinogenesis in the past. So far, we know that arsenic toxicity involves genetic and epigenetic changes, and arsenic can affect reactive oxygen production and oxidative stress, enhanced cell proliferation and modulation of gene expression, reduce DNA repair, increase growth factors, and alter DNA methylation and signal transduction, but the molecular mechanism of arsenic-induced human skin and internal organs carcinogenesis are not completely understood. In

* Corresponding author: Ning Ma (MD., PhD. Professor), Faulty of Health Science, Suzuka University of Medical Science, Suzuka, Mie 510-0293, Japan. Email: maning@suzuka-u.ac.jp.

the present review, we have attempted to evaluate and update the carcinogenesis mechanism of arsenic and its compounds based on our research in malignant transformation models in vitro and the available literature.

INTRODUCTION

Inorganic arsenic is clearly considered a human carcinogen by the World Health Organization and the International Agency for Research on Cancer (Mann et al., 2008). A large number of epidemiological data indicates that environmental exposure or occupational exposure to arsenic can cause skin, bladder and lung cancer. Liver and prostate tissues are now potentially considered to be potential human targets of arsenic as well (Alonso et al., 2010; Celik et al., 2008; Chen et al., 2010; Meliker et al., 2010; Morales et al., 2000; Smith et al., 1992). Furthermore, diabetes, cardiovascular disease and some nervous system dysfunction are related to arsenic exposure (Chen et al., 2007; Tseng et al., 2000). Repeatable evidence of complete carcinogenic effects after inorganic arsenic exposure was not available in animals, hindering mechanistic studies of arsenic carcinogenesis in the past. Some studies showed that experimental animals are likely to be much less sensitive to arsenic than humans; experimental animals often require 10 to 100 times the dose observed as having the same toxic effects on humans (Carter et al., 2003). Arsenic carcinogenic animal models are likely not available due to mature cells that are not sensitive to arsenic toxicity. Generally, the early stage of life can be a period of high sensitivity to carcinogenesis induced by many chemicals, because this period of life is involved in a variety of cell proliferation, differentiation, apoptosis and other biological processes. A more important factor is related to the large number of stem cells (SCs) in this process (Anderson, 2004; Anderson et al., 2000; Birnbaum and Fenton, 2003; Waalkes et al., 2007). In the present review, we have attempted to evaluate and update the carcinogenesis mechanism of arsenic and its compounds based on our research in malignant transformation models in vitro and available literature.

THE RELATIONSHIP BETWEEN CANCER AND STEM CELLS

There was an assumption that cancers may develop from stem cells, which originated over 100 years ago (Polyak and Hahn, 2006). So far, the existence of cancer stem cells in the cancer cell culture, leukemia and solid tumors remain to be further studied. But research evidence suggests that cancer stem cells are present in the majority of tumor cell population. Cancer stem cells have been reported including "stem-like cells" or "tumor initiating cells" in human solid tumors, such as brain tumors (Hemmati et al., 2003; Singh et al., 2004a; Singh et al., 2003; Singh et al., 2004b), prostate cancer (Al-Hajj et al., 2003; Collins et al., 2005; Richardson et al., 2004), breast cancer (Al-Hajj et al., 2003; Ponti et al., 2005), ependynoma (Taylor et al., 2005), skin cancer (Perez-Losada and Balmain, 2003), retinal neuroblastoma (Seigel et al., 2005), pancreatic cancer (Li and Yang, 2007), and hepatoblastoma (Fiegel et al., 2004) etc. Studies in vivo reported that cancer stem cells exist in stomach (Houghton et al., 2004) and lung cancer (Kim et al., 2005). Many studies in vitro also found cancer stem cells persist in a variety of tumor cell lines; all these studies suggest that cancer is likely to be a stem cell 'disease' (Kondo et al., 2004; Patrawala et al., 2005; Setoguchi et al., 2004).

Stem cells are a class of self-replication (self-renewing) pluripotent cells, which can differentiate into a variety of cell functions generally divided into embryonic stem cells (embryonic stem cell, ES cell), bone marrow stem cells (bone marrow derived stem cell, BMSC) or hematopoietic stem cells (hematopoietic stem cell, HSC) and adult stem cells (somatic stem cell) under certain conditions. Virtually all tissues and organs present with a small group of stem cells; the most obvious difference between stem cells and other cells is the stem cells' self-renewal capacity (Pardal et al., 2003; Polyak and Hahn, 2006; Visvader and Lindeman, 2008; Wicha et al., 2006). One stem cell can divide into two cells, but one of the daughter cells remains in an undifferentiated state identical to the parent cell while the other daughter cell is directed to differentiate. Stem cells can be identified by the expression of surface molecules, such as CD133, CD34, CD44, CD117, etc. (Bomken et al., 2010), and the cell self-renewal-related genes were highly expressed, such as Oct-4, p63, Bmi-1, Wnt, Notch (Lobo et al., 2007; Reya et al., 2001; Visvader and Lindeman, 2008). Cancer stem cells with stem cell-like properties are a small fraction of the population of cancer cells, cells that have the potential for cancer formation, especially metastatic cancer. Cancer stem cells plays a key rope in the spread of a cancer from one organ or part of an organ to another non-adjacent organ or part. The "tumor-like" stem cells can be found in many tumor tissues, and cancer stem cells may be derived from the differentiation or proliferation of normal stem cells (Bomken et al., 2010; Lobo et al., 2007; Wicha et al., 2006). Cancer stem cells and normal stem cells are similar in some of the key features of morphological and cell membrane markers. Stem cells and cancer stem cells have a similar signal to regulate the growth of pathways (Lobo et al., 2007; O'Brien et al., 2010; Pardal et al., 2003; Reya et al., 2001; Visvader and Lindeman, 2008). In the transfer process of normal cells to cancer cells, the self-renewing and long-term survival in normal tissue stem cells may be the accumulation of the "multiple attack" and eventually become immortal cancer stem cells. Therefore, existing studies suggest that tumors may be derived from the accumulation of the repeated stimulation of carcinogens and the malignant transformation of normal stem cells.

THE MECHANISM OF ARSENIC-INDUCED CANCER

Chronic arsenic exposure is associated with high risk for skin, lung and liver cancers (Alonso et al., 2010; Celik et al., 2008; Chen et al., 2010; Meliker et al., 2010; Morales et al., 2000; Smith et al., 1992). Normal tissue cell characteristics immortalized human keratinocyte cell lines (HaCaT), human embryonic lung fibroblast cell lines (HELF), small airway epithelial cells (SAECs), and liver epithelial cells. TRL 1215 cells are commonly used in arsenide cell malignant transformation and its mechanism in vitro studies. Although the exact mechanisms remain to be studied, potential carcinogenic actions for arsenic in malignant transformation model studies in vitro include regulation of cell signaling pathways, oxidative stress-induced DNA damage, arsenic-induced chromosomal abnormalities, growth factor changes, cell proliferation, DNA repair change, inhibition of tumor suppressor genes, DNA methylation pattern changes, gene amplification and other mechanisms. Studies have shown that arsenic exposure can activate multiple signaling pathways, including ATM, *p38*, MAPK, *p53*, Notch, IFN-â, Nrf2-Keap1 antioxidant, Wnt signaling pathways (Menendez et al., 2001; Takahashi et al., 2013; Xia et al., 2012; Yoda et al., 2008; Zhang et al., 2011), and inhibit Signal Transducers and Activator of Transcription (STAT) signaling pathway (Hayashi et al., 2002). Much of the in vitro work has been deeply researched on some of these pathways; MAPKs signaling family pathways have an important regulation role on cell proliferation and differentiation. Extra-cellular signal-regulated protein kinases (Erks), *p38* and c-Jun NH2-terminal kinases (JNK) were sensitive to the arsenic toxicity and very important in arsenic-induced malignant transformation. Studies showed that arsenite induces phosphorylation of Erks and JNKs, arsenite-induced Erk activation was markedly inhibited by introduction of dominant-negative Erk2 into cells. The activation of Erks is required for arsenic-induced cell transformation, whereas the activation of JNKs and NF-κB is involved in arsenic-induced apoptosis of JB6 cells (Dong, 2002; He et al., 2007; Lau et al., 2013; Li and Yang, 2007; Pei et al., 2008; Qu et al., 2002; Thorsen et al., 2006). NF-κB (nuclear factor kappa-light-chain-enhancer of activated B cells) is a protein complex that controls DNA transcription, regulates the expression of apoptosis and proliferation-related genes, and plays an important role in the carcinogenesis process (Sakon et al., 2003). Incorrect regulation of NF-κB has been linked to cancer, inflammatory and autoimmune diseases, etc. Cytokines, a variety of environmental and occupational hazards, metal poisoning, intracellular stress, UV and mitogenic agents can stimulate the activation of NF-κB factor (Karin, 2006). But the findings of the arsenic affect signaling transduction of NF-κB is not consistent; high concentrations of inorganic arsenic may inhibit the activation of NF-κB signaling pathway in a variety of cell systems, and often induce their apoptosis signal closely related to the conduction. In the low concentration experiment, arsenic is observed to enhance NF-κB activation and DNA binding ability, and is often associated with cell proliferation but not apoptosis. Different concentrations of arsenic may cause different effects (Bode and Dong, 2002). Many studies have shown that arsenic induced DNA damage is caused by triggering ATM/MDM2/p53 signaling pathways and cell apoptosis (Jomova et al., 2011; Platanias, 2009). The Akt pathway is also affected by arsenic toxicity in alteration of its multiple pathways to coordinate and decide the cell proliferation or apoptosis (Chen et al., 2012; Mann et al., 2008). Arsenic can inhibit the cell gap junction channel (GJIC) and GJIC plays an important role in promoting the growth stage of cancer (Deng et al., 2002). A human investigation study showed that arsenic toxicity and carcinogenesis may be associated with multiple signaling pathways related to arsenic exposure. Chronic arsenic exposure induced the activation of NF-κB and the interleukin-1 (IL-1) signaling pathway, which plays an important role in chronic arsenic poisoning (Argos et al., 2006; Fry et al., 2007). Overall, results of the current study about arsenic's effect on signal transduction are difficult to explain, purely on one or two pathways of arsenic's mechanism of action. Cross-linking multiple pathways (cross talk) will greatly increase the difficulty of the study.

Many arsenic carcinogenicity studies have shown that arsenic exposure can induce reactive oxygen species (ROS) to increase (Shi et al.,

2004). Arsenic alters mitochondrial membrane integrity and causes mitochondrial membrane dysfunction. Altered mitochondrial functions are considered to increase the formation of superoxide anion radical generating source, causing glutathione depletion and promoting cell sensitivity to arsenic toxicity (Cohen et al., 2006; Valko et al., 2005). Many studies on human and animal experiments show that arsenic exposure may increase the formation of peroxide radicals (ROO•), superoxide anion (O2•-), hydroxyl radicals (•OH), and the blood of non-protein thiol (Flora et al., 2007). These radicals are considered among the most critical factors effecting DNA strand breaks or damage. The oxidation of hemoglobin is one of the main ways to generate superoxide radicals; arsenic can contribute to increased production of heme oxygenase, which is a reason for the occurrence of oxidative stress caused by arsenic in vitro studies (Applegate et al., 1991). Oxygen free radical scavenger dimethyl sulfoxide (DMSO) intervention can significantly reduce the mutagenicity of arsenic (Hei et al., 1998). The study also found that arsenic exposure could cause dose-depend radicals to increase in the culture cell. When adding catalase (CAT) and catalytic superoxide dismutase (SOD), it can reduce the amount of free radicals significantly in the cells (Liu et al., 2001). Arsenic induced hydrogen peroxide and superoxide radicals snatched on protons cause DNA damage and changes in the structure of DNA, leading to cell mutagenic, carcinogenic or cell death. 8-Hydroxy-deoxyguanosine is one of the major reactive oxygen species-induced DNA base-modified products that is widely accepted as a sensitive marker of oxidative DNA damage. Experiments on humans and animals indicate that 8-OHdG increases in the tissues of skin cancer induced by exposure to arsenic. The involvement or reactive oxygen species in arsenic-induced human skin cancer is strongly suggested (Matsui et al., 1999; Vijayaraghavan et al., 2001; Wanibuchi et al., 1997). Exposure to arsenic-induced reactive oxygen species not only causes DNA damage, it can also cause DNA strand breaks, including single-stranded and double-strand breaks. Many studies showed that dimethylarsinic acid (DMAV) could cause single-stranded DNA and double-strand breaks. In the alkaline single-cell gel electrophoresis (comet assay, SCGE), it was found that arsenic can be generated ROS and induce DNA damage in a variety of human cells. Adding oxygen free radical scavenger DMSO and CAT can protect or mitigate the damage to cells (Li et al., 2001). The production of reactive oxygen species is mainly from the dimethylarsinic acid DMAV metabolism, which causes DNA damage in arsenic treated rats (Yamanaka et al., 1991). However, some researchers consider that arsenic did not directly cause

DNA strand breaks; they deem it as a weak breaking agent (Brown and Kitchin, 1996).

Arsenic exposure can cause human skin keratinocyte proliferation, make mRNA transcription enhancers, transform growth factor-α (TGF-α), granulocyte-macrophage colony-stimulating factor (GM-CSF) and cause tumor necrosis factor-α (TNF-α) to increase (Germolec et al., 1997). Arsenic exposure also increases the intracellular ROS levels by activating transcription factors (such as AP-1, c-fos, and NF-κB), so that the immunogenicity and promotion of excessive expression of inflammatory cytokines, leading to excessive cell proliferation, can ultimately cause cancer (Kitchin, 2001). Arsenic exposure can also change DNA repair; arsenic inhibits more than 200 kinds of known enzymes (Abernathy et al., 1999). Monomethylarsonous acid (MMA III), dimethylarsinous acid (DMA III) and trimethylarsenite (TMA III) in the arsenite can easily bind to dimethylarsinic glutathione protein, causing protein conformational changes and lead to inactivation of enzymes (Kitchin, 2001). Trivalent arsenic has a powerful combination with free sulfur; this combination can lead to inhibition of DNA repair, key genetic locus mutation and cells proliferation (Kitchin, 2001). Arsenic carcinogenesis also includes the effect of inhibition of tumor suppressor genes. Arsenite reduced p53 levels while concomitantly increasing the p53 regulatory protein mdm2 levels in a dose- and time-dependent manner (Hamadeh et al., 1999; Salazar et al., 1997).

STEM CELLS AND ARSENIC CARCINOGENICITY

Stem cells are a class of self-renewal during the development of the individual, highly proliferative and differentiation potential of the cell population. These cells maintain their division and further differentiate into various tissues to form complex tissues and organs of the body. Stem cells maintain normal physiological function of tissues, and organs mature through the proliferation and differentiation of cells. They also play an important role in tissue repair and organ damage. Cancer stem cells in tumors have similar self-renewing, highly proliferative characteristics as seen in stem cells (Bomken et al., 2010; Lobo et al., 2007; Pardal et al., 2003; Visvader and Lindeman, 2008). Many current studies have shown that a population of stem cells with abnormal activation of cell self-renewal, differentiation, and proliferation-related signaling pathways is able to undergo self-renewing divisions and differentiate to form all populations within a malignancy. Since cancer stem cells and stem cells have similar signaling pathways to regulate growth features, the tumor can be considered to be a small group of self-cells (cancer stem cells) composed of abnormal tissue renewal and proliferation, but the tissue cells value is rendered to differentiation disorder, no self-stability tends to malignant immortalized. (Bomken et al., 2010; Lobo et al., 2007; Pardal et al., 2003; Reya et al., 2001).

The skin is a major target of arsenic exposure and one of the most sensitive tissues to chronic arsenic exposure. Chronic exposure to arsenic results in various epidermal alterations, including hyperpigmentation, hyperkeratosis, squamous cell carcinoma (SCC) and basal cell carcinoma. Although inorganic arsenic alone does not seem to independently induce skin cancer in animals, it definitely acts together with UV irradiation to stimulate skin cancer in mouse (Rossman et al., 2001; Rossman et al., 2002). Nevertheless, there is still a lack of specific molecular biomarkers to detect arsenic-induced early malignant lesions. Increasing recent evidence indicates that cancers may develop from a small subpopulation of CSCs present in both solid tumors and hematological malignancies (Lobo et al., 2007; O'Brien et al., 2010; Pardal et al., 2003; Polyak and Hahn, 2006; Visvader and Lindeman, 2008). Analysis of CSCs biomarker expression during tumor formation may help to gain a better insight into the process of tumorigenesis and to identify new markers for early diagnosis of malignant transformation (Visvader and Lindeman, 2008). The CSCs biomarkers can be identified by the expression of a variety of cell surface markers, including CD133 and CD44, some of which have phenotypic significance (Bomken et al., 2010) and show higher expression of a variety of stem cell maintenance-related genes, including Oct-4, p63, and Notch (Lobo et al., 2007; Reya et al., 2001; Visvader and Lindeman, 2008). Recent studies have reported that arsenic and its compounds can affect a variety of cell lines and normal stem cells in vivo kinetics of proliferation and cell differentiation (Patterson et al., 2005; Patterson and Rice, 2007; Tokar et al., 2010a; Tokar et al., 2010d; Waalkes et al., 2008). Arsenite-treated cultures exhibited elevated levels of beta1-integrin and beta-catenin, two proteins enriched in cells with high proliferative potential. These findings suggest that arsenic could have co-carcinogenic and tumor co-promoting activities in the epidermis as a result of increasing the population and persistence of germinative cells targeted by tumor initiators and promoters (Patterson et al., 2005; Patterson and Rice, 2007). In vitro, arsenic has been shown to have an ability to snare normal stem cells, and gradually induce the change of normal stem cells into cancer-like stem cells (Tokar et al., 2010c; Tokar et al., 2010d). Chronic treatment of arsenic induced self-renewal of normal stem cells and up/down-regulation related genes include the own-regulation of PTEN (Tokar et al., 2010c). Treatment of arsenic on the normal stem cells under a low concentration, cause the populations of cancer stem cells to increase, thus greatly improving the probability of malignant transformation. It also suggests that if arsenic exposure occurs during the stem cell's proliferation or differentiation active periods, such as in the prenatal period and/or early developmental stages of life, it will lead to increase the risk of carcinogenesis. This also explains why in utero exposure to arsenic can induce cancer in the offspring in adulthood (Tokar et al., 2010b; Waalkes et al., 2006a; Waalkes et al., 2006b; Waalkes et al., 2004; Waalkes et al., 2003), showing a minimum risk of arsenic exposure to the most adult organs in rodents (Tokar et al., 2010a). Consistent with previous studies, we found that chronic exposure to low concentration of 0.05 ppm of arsenic, morphological differences were observed between the arsenic-treated HaCaT cells and passage-matched control cells. Whereas control cells maintained an epithelial-like morphology, arsenic-treated cells developed into giant multinuclear cells, and exhibited a more vigorous growth capacity (Figure 1). 8-OHdG formation was found in the 0.05 ppm arsenic exposure HaCaT cells from 2 weeks (5 passages), and showed time course increasing intense staining in the cell nucleus (Figure 2). The incidence of colony formation was increased during arsenic exposure period, which after 30 passages of arsenic exposure was 4.2-fold higher in the arsenic-treated cells than in passage-matched control cells. And the changes in the cell cycle showed that the majority of cells were in the G2 and S phase. Moreover, arsenite-treated cells rapidly developed into highly pleomorphic tumors, with regional invasion in nude mice. The skin tumors were highly undifferentiated, highly malignant, and composed of immature epithelial- and mesenchymal-like cells, and were identified as squamous cell carcinomas (SCC) (Figure 3). These characteristics show that arsenic-treated HaCaT cells rapidly acquired a cancer phenotype upon exposure to arsenite (Huang et al., 2013). The levels of CD44v6 and CD133 proteins of HaCaT cells exposed to 0.05 ppm and 0.1 ppm of arsenic were analyzed at different post-exposure times by immunofluorescence histochemistry. Figure 4 shows that CD44v6 protein was increased to a significant extent in the arsenic-treated group from 25 passages relative to HaCaT cells in the 0 ppm arsenic group. Immunohistochemistry showed that CD44v6 was strongly stained in the cell membrane of arsenic-treated HaCaT cells after 25 passages (Figure 4. As 0.05 ppm, 0.1 ppm). Control cells were showing either no or weak immunoreactivity (Figure 4. As 0 ppm). In the arsenic-treated group, quantitative analysis of CD44v6 immunoreactive HaCaT cells showed that the immunoreactivity increased proportionally to the number of passages from 15 to 25 passages. A long-term exposure of HaCaT cells to arsenic caused the activity of CD44v6 to increase with an increase in the duration of the exposure, that the extent of CD44v6 immunoreactivity in arsenic-treated cells was found to positively correlate with the cloning efficiency in soft agar (r=0.949, P=0.01). On the other hand, weak immunoreactivity for CD133 was observed in arsenic-treated cells, and no immunoreactivity was observed in passage-matched control cells. The expression of the specific cell adhesion molecule CD44v6 has been shown to be associated with metastasis and poor prognosis in certain human malignancies, such as breast cancer and colorectal cancer (Liu et al., 2005). Increased levels of CD44 and/or different patterns of splice variants were found in tumors in comparison with their normal counterparts (Herrlich et al., 1995; Stauder et al., 1995). In our study, we found CD44v6 was the independent biological prognostic marker for the formation colony in soft agar of arsenic treated cells, and CD44v6 was strongly expressed in tumor cells induced by arsenic (Figure 5), suggesting CD44v6 is a valuable molecular marker for various human skin lesions, including squamous cell carcinoma (SCC) that results in chronic arsenic exposure.

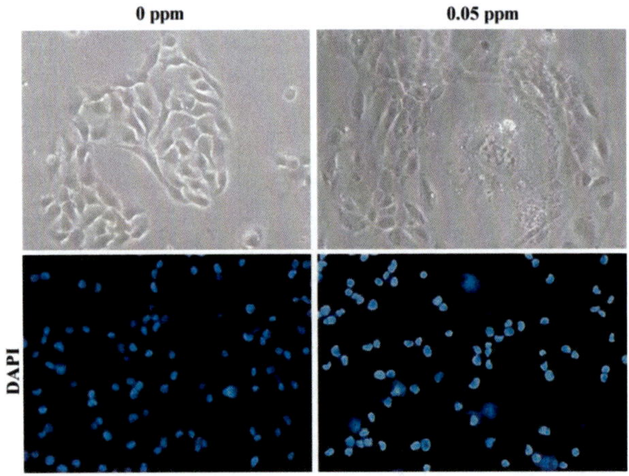

Figure 1. Low-level, chronic arsenic exposure induces malignant transformation in HaCaT cells. Control cells (0 ppm) maintained an epithelial-like morphology at 25 passages stage, arsenic-treated cells exhibited morphological alterations with the frequent occurrence of giant multinuclear cells at 25 passages stage (200×magnification) (0.05 ppm). Control HaCaT cells and arsenic-treated cells at 25 passages stage stained with DAPI. Arrows indicate giant multinuclear cells.

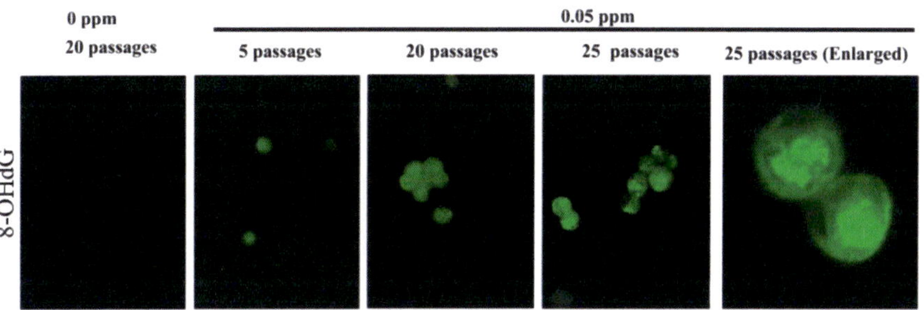

Figure 2. 8-OHdG immunoreactivity is localized in arsenic-treated HaCaT cells. HaCaT cells were exposed to 0.0 ppm and 0.05 ppm sodium arsenite from 0 passages to 25 passages. Immunofluorescence immunoreactivity of 8-oxodG was observed mostly in the cell nucleus and the staining intensity was greatest on 25 passages.

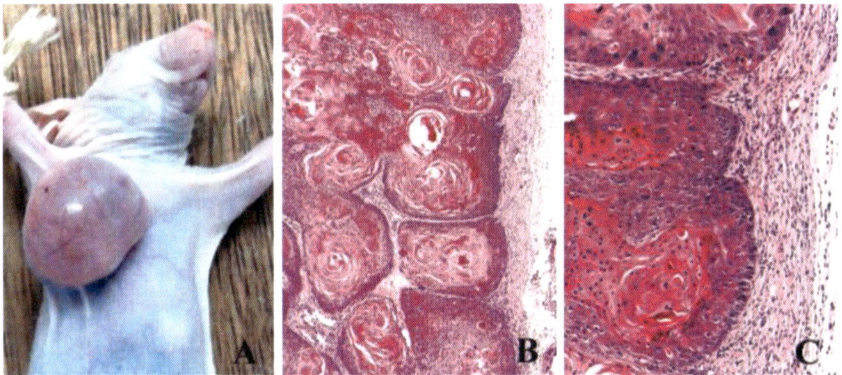

Figure 3. Arsenic-treated 25 passages cells (2×106/0.2 mL) were injected subcutaneously on the left axillary fossa of Balb/c nude mice. (A) Arsenic-treated HaCaT cells became apparently tumorigenic in nude mice. (B), (C) Pathological studies of Xenograft tumor sections showed a highly undifferentiated, pleomorphic nature of tumors from arsenic-treated HaCaT cells, and the tumor testified as squamous cell carcinomas (SCC), a type of cancer induced by arsenic in skin.

In vivo studies, fetal exposure to arsenic in mice can induce or initiate tumors or preneoplasias in a wide variety of tissues in adulthood. Pregnant mice were treated with drinking water containing sodium arsenite at up to 85ppm arsenic from days 8 to 18 of gestation, and the offspring were observed for up to 2 years. Male CD1 mice treated with arsenic in utero develop tumors of the liver and adrenal and renal hyperplasia while females develop tumors of urogenital system, ovary, uterus and adrenal and hyperplasia of the oviduct (Waalkes et al., 2007), including several concordant with human sites of arsenic carcinogenesis, like the bladder and lung (Celik et al., 2008; Waalkes et al., 2007). The tendency for arsenic to act as a transplacental carcinogen seems to be in taking with fetal stem cells as a critical target in carcinogenesis. The occurrence of skin cancer was observed in their adulthood of perinatal arsenic exposure in utero CD1 rats. Cancer stem cell marker CD34-positive cells observed a significant increase in this cancer tissue. Fetal arsenic exposure facilitated a cancer response in association with distorted skin tumor stem cell signaling and population dynamics, implicating stem cells as a target of arsenic in the fetal basis of skin cancer in adulthood (Waalkes et al., 2008). Long-term exposure to low concentrations NaAsO2 can directly induce normal prostate epithelial stem cells into tumor-like stem cells (cancer stem cells) (Tokar et al., 2010a; Tokar et al., 2010d). In the arsenic-induced normal stem cells into cancer stem cell-like process, self-renewal-related genes *p63*, ABCG2, BMI-1, sonic hedgehog, OCT-4, and NOTCH-1 express a first down- and then up-regulated pattern, the expression of those genes showed alterations in a U-shape. These self-renewal gene expression changes were very similar to the development of gene expression changes in the cancer stem cells (Lobo et al., 2007; Pardal et al., 2003; Reya et al., 2001; Visvader and Lindeman, 2008), and the self-renewal of chronic arsenic treatment resulted in a U-shaped related gene expression was also observed in the Krivtsov et al. study (Krivtsov et al., 2006). Phosphatase and tensin homologue deleted on chromosome ten (PTEN) was depleted to low concentrations on the epithelial normal stem cell in the chronic arsenic exposure. PTEN gradual down regulation may play an important role in this process because its depletion increases the normal stem cell self-renewal related genes expression. The tumor tissue stem-like cells increased significantly, resulting in cancer tissues and cancer stem cell enrichment (Dubrovska et al., 2009; Groszer et al., 2006; Yilmaz et al., 2006). This data strongly suggests that the PTEN/PI3K/Akt pathways are critical for maintenance of cancer stem-like cells and that targeting PI3K signaling may be beneficial in cancer treatment by eliminating cancer stem-like cells.

Arsenic-induced malignant transformation from normal stem cells may be due to the innate resistance to arsenic as a metalloid and protoplasm poison. It has a higher affinity than other metals (Tokar et al., 2010b). This inherent resistance of stem cells and cancer stem cells make the cells have a high affinity to arsenic and induces malignant transformations to occur gradually. In the other carcinogenic agents such as cadmium, N-methyl-n-nitrosourea (MNU)-induced malignant transformation process is not observed this phenomenon (Tokar et al., 2010b), indicating that this phenomenon may be a special mechanism of arsenic carcinogenicity. Arsenic and cadmium co-administration, to induce malignant transformation epithelial keratinocytes, resulting in a malignant transformation process was observed (Patterson et al., 2005).

Stem cells are appealing candidates as the cell of origin for cancer because of their pre-existing capacity for self-renewal and unlimited replication (Sell, 2006). Normal and cancer SCs exist in two hierarchical subpopulations: quiescent and proliferating. Absolute numbers of quiescent SCs are comparable in normal and cancer SCs, however, normal SCs are mainly quiescent, whereas CSCs are mainly proliferating, a case in point is chronic tissue repair, that aberrant alterations in self-renewal pathways may prevent activated SCs from returning to a quiescent state that usually follows regeneration, thereby enhancing the probability of an oncogenic event during chronic injury (Beachy et al., 2004). Recent research evidence suggests that inorganic arsenic in drinking water ingestion in early childhood might increase the risks of childhood liver cancer mortality (Liaw et al., 2008; Marshall et al., 2007; Smith et al., 2006; Yuan et al., 2010). The major route of arsenic exposure includes drinking arsenic contaminated water from the environment (Liaw et al., 2008; Matschullat, 2000; Smith et al., 2006; Yuan et al., 2010), and ingestion of arsenic due to contamination of food (Yorifuji et al., 2010). Transplacental arsenic exposure in the animal model experiment study showed that arsenic administration from drinking water to pregnant mice in the early pregnant rat, the offspring developed tumors in the urogenital system, ovary, uterus and adrenal (Tokar et al., 2010b; Waalkes et al., 2006a; Waalkes et al., 2006b; Waalkes et al., 2004; Waalkes et al., 2003). Together, those human data and rodent experimental observations hint that early life exposure to environmental arsenic can cause carcinogenesis in human beings. In vivo studies show that cancer stem cells may play a key role in the process of carcinogenesis. An apparent stem cell survival advantage with regard to arsenic causes selection during malignant transformation that manifests itself as an overabundance of cancer stem cells specifically after arsenic-driven acquisition of malignant phenotype (Patterson et al., 2005; Patterson and Rice, 2007; Tokar et al., 2010c; Tokar et al., 2010d).

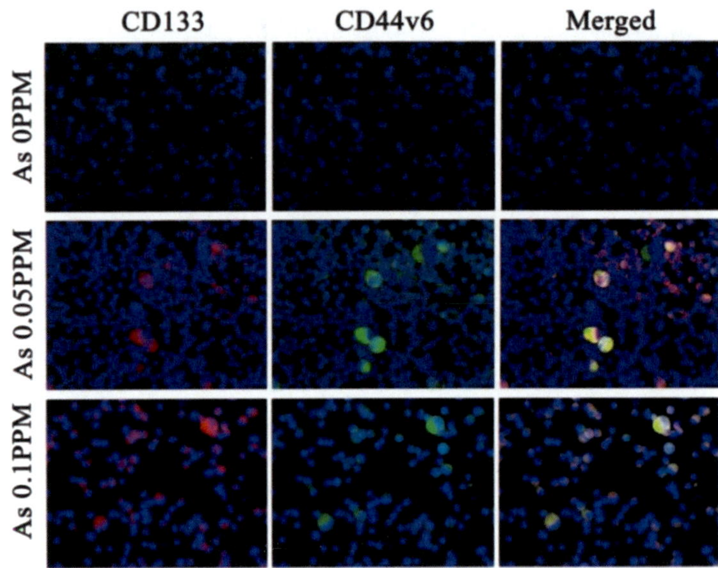

Figure 4. CD44v6 and CD133 proteins of HaCaT cells exposed to 0.05 ppm and 0.1 ppm of arsenic were analyzed at 25 passages post-exposure times by immunofluorescence histochemistry.

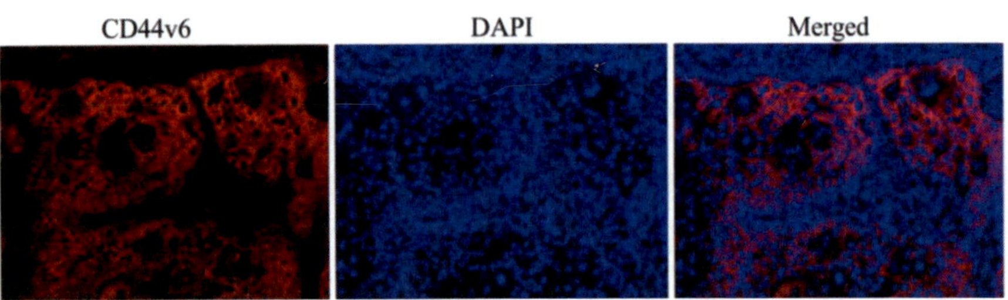

Figure 5. CD44v6 expression in tumor cells of Xenograft tumor. CD44v6 expression was estimated by the percentage of tumor cell membrane staining, exceed 50% tumor cells were CD44v6 immunoreactivity positive.

THE OUTLOOK

Studies of arsenic pathogenesis in animal models have been difficult to carry out so far. It was often conducted by epidemiological studies. Arsenic-induced malignant transformation on cells, and arsenic carcinogenicity as a possible mechanism, provides a lot of useful information in vitro experiments.

However, many of the limitations of these experiments should be recognized; in vitro reaction system by a variety of factors including exposure dose, exposure time, the length of the observation period, culture conditions and cell or tissue sources. As a metal material, arsenic has features similar to carbon. Arsenic is not easy to combine with other substances, and it has a higher variety valence of oxidation states. Some studies show that there are no completely arsenic compounds of carcinogenic mechanism that can be ruled out, and probably more than one pathway or mechanism is at work (Aposhian et al., 2004; Naranmandura et al., 2007; Suzuki et al., 2002; Wanibuchi et al., 2004). The possibility that inorganic arsenic exposure at fetal or infant stage has a close relation with the occurrence of cancer in adulthood has received special attention in recent years. As an environmental poison, the incubation period of arsenic carcinogenic to human beings is 30 to 50 years. Some of the current studies suggest that arsenic-induced tumors may be a stem cell disease; although there is little evidence for this theory, there are still experiments with convincing arguments. Perhaps the early life of the stem cells is a major target of arsenic. Mutations in somatic cells do not cause tumors. They occur because of a short half-life of mature cell survival, an oncogenic progress before the incident, or two hits before death occurs naturally or they are killed. Cancer stem cells or precursor cells must be present in the tissues before the tumor formation, because these cells can be sustained for longer survival. Therefore, cancer stem cells may be derived from abnormal proliferation and self-renewal of normal stem cells, or obtained from self-refreshing precursor cells (Soltysova et al., 2005). Recent studies suggest that low-level arsenic-treated HaCaT cells may develop a typical skin cancer in which cancer stem cells exist with the capacity of self-renewal and potential differentiation. CD44v6 may be a biomarker of arsenic-induced neoplastic transformation in human skin cells, and arsenic promotes malignant transformation in HaCaT cells through its activation of NF-κB and inactivation of *P53* stimulated expression of CD44v6 (Huang et al., 2013).

Is arsenic carcinogenesis a stem cell disease? The key distinction is to find and identify the cancer stem cells in the arsenic-induced tumors, and other tumor-specific biomarker molecules in cancer stem cells. This will further promote a more comprehensive and in-depth study to understand the molecular mechanism of arsenic-induced carcinogenesis. A clear signal of its network and carcinogenic pathway target molecules, as well as valuable biomarkers and treatment methods for arsenide, will provide valuable information towards cancer prevention strategies.

REFERENCES

Abernathy, C. O., Liu, Y. P., Longfellow, D., Aposhian, H. V., Beck, B., Fowler, B., Goyer, R., Menzer, R., Rossman, T., Thompson, C. & Waalkes, M. (1999). Arsenic: health effects, mechanisms of actions & research issues. *Environmental health perspectives, 107*, 593-597.

Al-Hajj, M., Wicha, M. S., Benito-Hernandez, A., Morrison, S. J. & Clarke, M. F. (2003). Prospective identification of tumorigenic breast cancer cells. *Proc Natl Acad Sci U S A, 100*, 3983-3988.

Alonso, F. T., Garmendia, M. L. & Bogado, M. E. (2010). Increased skin cancer mortality in Chile beyond the effect of ageing: Temporal analysis 1990 to 2005. *Acta dermato-venereologica, 90*, 141-146.

Anderson, L. M. (2004). Predictive values of traditional animal bioassay studies for human perinatal carcinogenesis risk determination. Toxicology and applied pharmacology, *199*, 162-174.

Anderson, L. M., Diwan, B. A., Fear, N. T. & Roman, E. (2000). Critical windows of exposure for children's health: cancer in human epidemiological studies and neoplasms in experimental animal models. Environmental health perspectives, *108 Suppl, 3*, 573-594.

Aposhian, H. V., Zakharyan, R. A., Avram, M. D., Sampayo-Reyes, A. & Wollenberg, M. L. (2004). A review of the enzymology of arsenic metabolism and a new potential role of hydrogen peroxide in the detoxication of the trivalent arsenic species. *Toxicology and applied pharmacology, 198*, 327-335.

Applegate, L. A., Luscher, P. & Tyrrell, R. M. (1991). Induction of heme oxygenase: a general response to oxidant stress in cultured mammalian cells. *Cancer research, 51*, 974-978.

Argos, M., Kibriya, M. G., Parvez, F., Jasmine, F., Rakibuz-Zaman, M. & Ahsan, H. (2006). Gene expression profiles in peripheral lymphocytes by arsenic exposure and skin lesion status in a Bangladeshi population. Cancer epidemiology, biomarkers & prevention: a publication of the American Association for Cancer Research, cosponsored by the American Society of Preventive Oncology, *15*, 1367-1375.

Beachy, P. A., Karhadkar, S. S. & Berman, D. M. (2004). Tissue repair and stem cell renewal in carcinogenesis. *Nature, 432*, 324-331.

Birnbaum, L. S. & Fenton, S. E. (2003). Cancer and developmental exposure to endocrine disruptors. *Environmental health perspectives, 111*, 389-394.

Bode, A. M. & Dong, Z. (2002). The paradox of arsenic: molecular mechanisms of cell transformation and chemotherapeutic effects. *Critical reviews in oncology/hematology, 42,* 5-24.

Bomken, S., Fiser, K., Heidenreich, O. & Vormoor, J. (2010). Understanding the cancer stem cell. *British journal of cancer, 103,* 439-445.

Brown, J. L. & Kitchin, K. T. (1996). Arsenite, but not cadmium, induces ornithine decarboxylase and heme oxygenase activity in rat liver: relevance to arsenic carcinogenesis. *Cancer letters, 98,* 227-231.

Carter, D. E., Aposhian, H. V. & Gandolfi, A. J. (2003). The metabolism of inorganic arsenic oxides, gallium arsenide, and arsine: a toxicochemical review. *Toxicology and applied pharmacology, 193,* 309-334.

Celik, I., Gallicchio, L., Boyd, K., Lam, T. K., Matanoski, G., Tao, X., Shiels, M., Hammond, E., Chen, L., Robinson, K. A., et al. (2008). Arsenic in drinking water and lung cancer: a systematic review. *Environmental research, 108,* 48-55.

Chen, B., Liu, J., Chang, Q., Beezhold, K., Lu, Y. & Chen, F. (2012). JNK and STAT3 signaling pathways converge on Akt-mediated phosphorylation of EZH2 in bronchial epithelial cells induced by arsenic. *Cell cycle, 12,* 112-121.

Chen, C. J., Wang, S. L., Chiou, J. M., Tseng, C. H., Chiou, H. Y., Hsueh, Y. M., Chen, S. Y., Wu, M. M. & Lai, M. S. (2007). Arsenic and diabetes and hypertension in human populations: a review. *Toxicology and applied pharmacology, 222,* 298-304.

Chen, C. L., Chiou, H. Y., Hsu, L. I., Hsueh, Y. M., Wu, M. M., Wang, Y. H. & Chen, C. J. (2010). Arsenic in drinking water and risk of urinary tract cancer: a follow-up study from northeastern Taiwan. Cancer epidemiology, biomarkers & prevention: a publication of the American Association for Cancer Research, cosponsored by the American Society of Preventive Oncology, *19,* 101-110.

Cohen, S. M., Arnold, L. L., Eldan, M., Lewis, A. S. & Beck, B. D. (2006). Methylated arsenicals: The implications of metabolism and carcinogenicity studies in rodents to human risk assessment. *Crit Rev Toxicol, 36,* 99-133.

Collins, A. T., Berry, P. A., Hyde, C., Stower, M. J. & Maitland, N. J. (2005). Prospective identification of tumorigenic prostate cancer stem cells. *Cancer research, 65,* 10946-10951.

Deng, F. R., Li, Y. H., Zhang, H. M. & Guo, X. B. (2002). Effect of arsenite and phenylarsine oxide on gap junctional intercellular communications between human skin fibroblasts. *China journal of pharmacology and toxicology*, 378-381.

Dong, Z. (2002). The molecular mechanisms of arsenic-induced cell transformation and apoptosis. *Environmental health perspectives 110 Suppl*, 5, 757-759.

Dubrovska, A., Kim, S., Salamone, R. J., Walker, J. R., Maira, S. M., Garcia-Echeverria, C., Schultz, P. G. & Reddy, V. A. (2009). The role of PTEN/Akt/PI3K signaling in the maintenance and viability of prostate cancer stem-like cell populations. *Proc Natl Acad Sci U S A*, *106*, 268-273.

Fiegel, H. C., Gluer, S., Roth, B., Rischewski, J., von Schweinitz, D., Ure, B., Lambrecht, W. & Kluth, D. (2004). Stem-like cells in human hepatoblastoma. The journal of histochemistry and cytochemistry: *official journal of the Histochemistry Society*, *52*, 1495-1501.

Flora, S. J. S., Bhadauria, S., Kannan, G. M. & Singh, N. (2007). Arsenic induced oxidative stress and the role of antioxidant supplementation during chelation: A review. *Journal of Environmental Biology*, *28*, 333-347.

Fry, R. C., Navasumrit, P., Valiathan, C., Svensson, J. P., Hogan, B. J., Luo, M., Bhattacharya, S., Kandjanapa, K., Soontararuks, S., Nookabkaew, S., et al. (2007). Activation of inflammation/NF-kappaB signaling in infants born to arsenic-exposed mothers. *PLoS genetics*, *3*, e207.

Germolec, D. R., Spalding, J., Boorman, G. A., Wilmer, J. L., Yoshida, T., Simeonova, P. P., Bruccoleri, A., Kayama, F., Gaido, K., Tennant, R., et al. (1997). Arsenic can mediate skin neoplasia by chronic stimulation of keratinocyte-derived growth factors. *Mutation research*, *386*, 209-218.

Groszer, M., Erickson, R., Scripture-Adams, D. D., Dougherty, J. D., Le Belle, J., Zack, J. A., Geschwind, D. H., Liu, X., Kornblum, H. I. & Wu, H. (2006). PTEN negatively regulates neural stem cell self-renewal by modulating G0-G1 cell cycle entry. *Proc Natl Acad Sci U S A*, *103*, 111-116.

Hamadeh, H. K., Vargas, M., Lee, E. & Menzel, D. B. (1999). Arsenic disrupts cellular levels of p53 and mdm2: a potential mechanism of carcinogenesis. *Biochemical and biophysical research communications*, *263*, 446-449.

Hayashi, T., Hideshima, T., Akiyama, M., Richardson, P., Schlossman, R. L., Chauhan, D., Munshi, N. C., Waxman, S. & Anderson, K. C. (2002). Arsenic trioxide inhibits growth of human multiple myeloma cells in the bone marrow microenvironment. *Molecular cancer therapeutics*, *1*, 851-860.

He, X. Q., Chen, R., Yang, P., Li, A. P., Zhou, J. W. & Liu, Q. Z. (2007). Biphasic effect of arsenite on cell proliferation and apoptosis is associated with the activation of JNK and ERK1/2 in human embryo lung fibroblast cells. *Toxicology and applied pharmacology, 220,* 18-24.

Hei, T. K., Liu, S. X. & Waldren, C. (1998). Mutagenicity of arsenic in mammalian cells: role of reactive oxygen species. *Proc Natl Acad Sci U S A, 95,* 8103-8107.

Hemmati, H. D., Nakano, I., Lazareff, J. A., Masterman-Smith, M., Geschwind, D. H., Bronner-Fraser, M. & Kornblum, H. I. (2003). Cancerous stem cells can arise from pediatric brain tumors. *P Natl Acad Sci USA, 100,* 15178-15183.

Herrlich, P., Pals, S. & Ponta, H. (1995). CD44 in colon cancer. *Eur J Cancer, 31A,* 1110-1112.

Houghton, J., Stoicov, C., Nomura, S., Rogers, A. B., Carlson, J., Li, H., Cai, X., Fox, J. G., Goldenring, J. R. & Wang, T. C. (2004). Gastric cancer originating from bone marrow-derived cells. *Science, 306,* 1568-1571.

Huang, S., Guo, S., Guo, F., Yang, Q., Xiao, X., Murata, M., Ohnishi, S., Kawanishi, S. & Ma, N. (2013). CD44v6 expression in human skin keratinocytes as a possible mechanism for carcinogenesis associated with chronic arsenic exposure. *Eur J Histochem, 57,* 1-9.

Jomova, K., Jenisova, Z., Feszterova, M., Baros, S., Liska, J., Hudecova, D., Rhodes, C. J. & Valko, M. (2011). Arsenic: toxicity, oxidative stress and human disease. Journal of applied toxicology: *JAT, 31,* 95-107.

Karin, M. (2006). NF-kappa B and cancer: Mechanisms and targets. *Mol Carcinogen, 45,* 355-361.

Kim, C. F., Jackson, E. L., Woolfenden, A. E., Lawrence, S., Babar, I., Vogel, S., Crowley, D., Bronson, R. T. & Jacks, T. (2005). Identification of bronchioalveolar stem cells in normal lung and lung cancer. *Cell, 121,* 823-835.

Kitchin, K. T. (2001). Recent advances in arsenic carcinogenesis: modes of action, animal model systems, and methylated arsenic metabolites. *Toxicology and applied pharmacology, 172,* 249-261.

Kondo, T., Setoguchi, T. & Taga, T. (2004). Persistence of a small subpopulation of cancer stem-like cells in the C6 glioma cell line. *Proc Natl Acad Sci U S A, 101,* 781-786.

Krivtsov, A. V., Twomey, D., Feng, Z., Stubbs, M. C., Wang, Y., Faber, J., Levine, J. E., Wang, J., Hahn, W. C., Gilliland, D. G., et al. (2006). Transformation from committed progenitor to leukaemia stem cell initiated by MLL-AF9. *Nature, 442,* 818-822.

Lau, A., Whitman, S. A., Jaramillo, M. C. & Zhang, D. D. (2013). Arsenic-mediated activation of the Nrf2-Keap1 antioxidant pathway. *Journal of biochemical and molecular toxicology, 27*, 99-105.

Li, J. P. & Yang, J. L. (2007). Cyclin B1 proteolysis via p38 MAPK signaling participates in G2 checkpoint elicited by arsenite. *Journal of cellular physiology, 212*, 481-488.

Li, S. D., Morimoto, K., Takeshita, T. & Lu, S. Q. (2001). Application of single cell gel electrophoresis (SCGE) assay: comparative study of DNA damage induced by arsenic in human cells. *Chinese journal of endemiology*, 55-59.

Liaw, J., Marshall, G., Yuan, Y., Ferreccio, C., Steinmaus, C. & Smith, A. H. (2008). Increased childhood liver cancer mortality and arsenic in drinking water in northern Chile. Cancer epidemiology, biomarkers & prevention: a publication of the American Association for Cancer Research, cosponsored by the American Society of Preventive Oncology, *17*, 1982-1987.

Liu, S. X., Athar, M., Lippai, I., Waldren, C. & Hei, T. K. (2001). Induction of oxyradicals by arsenic: implication for mechanism of genotoxicity. *Proc Natl Acad Sci U S A, 98*, 1643-1648.

Liu, Y. J., Yan, P. S., Li, J. & Jia, J. F. (2005). Expression and significance of CD44s, CD44v6, and nm23 mRNA in human cancer. *World J Gastroenterol, 11*, 6601-6606.

Lobo, N. A., Shimono, Y., Qian, D. & Clarke, M. F. (2007). The biology of cancer stem cells. *Annual review of cell and developmental biology, 23*, 675-699.

Mann, K. K., Colombo, M. & Miller, W. H., Jr. (2008). Arsenic trioxide decreases AKT protein in a caspase-dependent manner. *Molecular cancer therapeutics, 7*, 1680-1687.

Marshall, G., Ferreccio, C., Yuan, Y., Bates, M. N., Steinmaus, C., Selvin, S., Liaw, J. & Smith, A. H. (2007). Fifty-year study of lung and bladder cancer mortality in Chile related to arsenic in drinking water. *Journal of the National Cancer Institute, 99*, 920-928.

Matschullat, J. (2000). Arsenic in the geosphere--a review. *The Science of the total environment, 249*, 297-312.

Matsui, M., Nishigori, C., Toyokuni, S., Takada, J., Akaboshi, M., Ishikawa, M., Imamura, S. & Miyachi, Y. (1999). The role of oxidative DNA damage in human arsenic carcinogenesis: Detection of 8-hydroxy-2'-deoxyguanosine in arsenic-related Bowen's disease. *Journal of Investigative Dermatology, 113*, 26-31.

Meliker, J. R., Slotnick, M. J., AvRuskin, G. A., Schottenfeld, D., Jacquez, G. M., Wilson, M. L., Goovaerts, P., Franzblau, A. & Nriagu, J. O. (2010). Lifetime exposure to arsenic in drinking water and bladder cancer: a population-based case-control study in Michigan, USA. *Cancer causes & control: CCC, 21*, 745-757.

Menendez, D., Mora, G., Salazar, A. M. & Ostrosky-Wegman, P. (2001). ATM status confers sensitivity to arsenic cytotoxic effects. *Mutagenesis, 16*, 443-448.

Morales, K. H., Ryan, L., Kuo, T. L., Wu, M. M. & Chen, C. J. (2000). Risk of internal cancers from arsenic in drinking water. *Environmental health perspectives, 108*, 655-661.

Naranmandura, H., Suzuki, N., Iwata, K., Hirano, S. & Suzuki, K. T. (2007). Arsenic metabolism and thioarsenicals in hamsters and rats. *Chem Res Toxicol, 20*, 616-624.

O'Brien, C. A., Kreso, A. & Jamieson, C. H. (2010). Cancer stem cells and self-renewal. Clinical cancer research: *an official journal of the American Association for Cancer Research, 16*, 3113-3120.

Pardal, R., Clarke, M. F. & Morrison, S. J. (2003). Applying the principles of stem-cell biology to cancer. *Nature reviews Cancer, 3*, 895-902.

Patrawala, L., Calhoun, T., Schneider-Broussard, R., Zhou, J. J., Claypool, K. & Tang, D. G. (2005). Side population is enriched in tumorigenic, stem-like cancer cells, whereas ABCG2(+) and ABCG2(-) cancer cells are similarly tumorigenic. *Cancer research, 65*, 6207-6219.

Patterson, T. J., Reznikova, T. V., Phillips, M. A. & Rice, R. H. (2005). Arsenite maintains germinative state in cultured human epidermal cells. *Toxicology and applied pharmacology, 207*, 69-77.

Patterson, T. J. & Rice, R. H. (2007). Arsenite and insulin exhibit opposing effects on epidermal growth factor receptor and keratinocyte proliferative potential. *Toxicology and applied pharmacology, 221*, 119-128.

Pei, B., Wang, S. C., Guo, X. Y., Wang, J., Yang, G., Hang, H. Y. & Wu, L. J. (2008). Arsenite-induced germline apoptosis through a MAPK-dependent, p53-independent pathway in Caenorhabditis elegans. *Chem Res Toxicol, 21*, 1530-1535.

Perez-Losada, J. & Balmain, A. (2003). Stem-cell hierarchy in skin cancer. *Nature Reviews Cancer, 3*, 434-443.

Platanias, L. C. (2009). Biological responses to arsenic compounds. The Journal of biological chemistry, *284*, 18583-18587.

Polyak, K. & Hahn, W. C. (2006). Roots and stems: stem cells in cancer. *Nature medicine, 12*, 296-300.

Ponti, D., Costa, A., Zaffaroni, N., Pratesi, G., Petrangolini, G., Coradini, D., Pilotti, S., Pierotti, M. A. & Daidone, M. G. (2005). Isolation and in vitro propagation of tumorigenic breast cancer cells with stem/progenitor cell properties. *Cancer research, 65,* 5506-5511.

Qu, W., Bortner, C. D., Sakurai, T., Hobson, M. J. & Waalkes, M. P. (2002). Acquisition of apoptotic resistance in arsenic-induced malignant transformation: role of the JNK signal transduction pathway. *Carcinogenesis, 23,* 151-159.

Reya, T., Morrison, S. J., Clarke, M. F. & Weissman, I. L. (2001). Stem cells, cancer, and cancer stem cells. *Nature, 414,* 105-111.

Richardson, G. D., Robson, C. N., Lang, S. H., Neal, D. E., Maitland, N. J. & Collins, A. T. (2004). CD133, a novel marker for human prostatic epithelial stem cells. *Journal of cell science, 117,* 3539-3545.

Rossman, T. G., Uddin, A. N., Burns, F. J. & Bosland, M. C. (2001). Arsenite is a cocarcinogen with solar ultraviolet radiation for mouse skin: an animal model for arsenic carcinogenesis. *Toxicology and applied pharmacology, 176,* 64-71.

Rossman, T. G., Uddin, A. N., Burns, F. J. & Bosland, M. C. (2002). Arsenite cocarcinogenesis: an animal model derived from genetic toxicology studies. *Environmental health perspectives, 110 Suppl,* 5, 749-752.

Sakon, S., Xue, X., Takekawa, M., Sasazuki, T., Okazaki, T., Kojima, Y., Piao, J. H., Yagita, H., Okumura, K., Doi, T. & Nakano, H. (2003). NF-kappa B inhibits TNF-induced accumulation of ROS that mediate prolonged MAPK activation and necrotic cell death. *Embo J, 22,* 3898-3909.

Salazar, A. M., Ostrosky-Wegman, P., Menendez, D., Miranda, E., Garcia-Carranca, A. & Rojas, E. (1997). Induction of p53 protein expression by sodium arsenite. *Mutation research, 381,* 259-265.

Seigel, G. M., Campbell, L. M., Narayan, M. & Gonzalez-Fernandez, F. (2005). Cancer stem cell characteristics in retinoblastoma. *Molecular vision, 11,* 729-737.

Sell, S. (2006). Cancer stem cells and differentiation therapy. Tumour biology: *the journal of the International Society for Oncodevelopmental Biology and Medicine, 27,* 59-70.

Setoguchi, T., Taga, T. & Kondo, T. (2004). Cancer stem cells persist in many cancer cell lines. *Cell cycle, 3,* 414-415.

Shi, H. L., Shi, X. L. & Liu, K. J. (2004). Oxidative mechanism of arsenic toxicity and carcinogenesis. *Molecular and cellular biochemistry, 255,* 67-78.

Singh, S. K., Clarke, I. D., Hide, T. & Dirks, P. B. (2004a). Cancer stem cells in nervous system tumors. *Oncogene, 23*, 7267-7273.

Singh, S. K., Clarke, I. D., Terasaki, M., Bonn, V. E., Hawkins, C., Squire, J. & Dirks, P. B. (2003). Identification of a cancer stem cells in human brain tumors. *Cancer research, 63*, 5821-5828.

Singh, S. K., Hawkins, C., Clarke, I. D., Squire, J. A., Bayani, J., Hide, T., Henkelman, R. M., Cusimano, M. D. & Dirks, P. B. (2004b). Identification of human brain tumour initiating cells. *Nature, 432*, 396-401.

Smith, A. H., Hopenhayn-Rich, C., Bates, M. N., Goeden, H. M., Hertz-Picciotto, I., Duggan, H. M., Wood, R., Kosnett, M. J. & Smith, M. T. (1992). Cancer risks from arsenic in drinking water. *Environmental health perspectives, 97*, 259-267.

Smith, A. H., Marshall, G., Yuan, Y., Ferreccio, C., Liaw, J., von Ehrenstein, O., Steinmaus, C., Bates, M. N. & Selvin, S. (2006). Increased mortality from lung cancer and bronchiectasis in young adults after exposure to arsenic in utero and in early childhood. *Environmental health perspectives, 114*, 1293-1296.

Soltysova, A., Altanerova, V. & Altaner, C. (2005). Cancer stem cells. *Neoplasma, 52*, 435-440.

Stauder, R., Eisterer, W., Thaler, J. & Gunthert, U. (1995). CD44 variant isoforms in non-Hodgkin's lymphoma: a new independent prognostic factor. *Blood, 85*, 2885-2899.

Suzuki, K. T., Mandal, B. K. & Ogra, Y. (2002). Speciation of arsenic in body fluids. *Talanta, 58*, 111-119.

Takahashi, M., Ota, A., Karnan, S., Hossain, E., Konishi, Y., Damdindorj, L., Konishi, H., Yokochi, T., Nitta, M. & Hosokawa, Y. (2013). Arsenic trioxide prevents nitric oxide production in lipopolysaccharide-stimulated RAW 264.7 by inhibiting a TRIF-dependent pathway. *Cancer science, 104*, 165-170.

Taylor, M. D., Poppleton, H., Fuller, C., Su, X., Liu, Y., Jensen, P., Magdaleno, S., Dalton, J., Calabrese, C., Board, J., et al. (2005). Radial glia cells are candidate stem cells of ependymoma. *Cancer cell, 8*, 323-335.

Thorsen, M., Di, Y., Tangemo, C., Morillas, M., Ahmadpour, D., Van der Does, C., Wagner, A., Johansson, E., Boman, J., Posas, F., et al. (2006). The MAPK Hog1p modulates Fps1p-dependent arsenite uptake and tolerance in yeast. *Molecular biology of the cell, 17*, 4400-4410.

Tokar, E. J., Benbrahim-Tallaa, L., Ward, J. M., Lunn, R., Sams, R. L., 2nd & Waalkes, M. P. (2010a). Cancer in experimental animals exposed to arsenic and arsenic compounds. *Crit Rev Toxicol, 40*, 912-927.

Tokar, E. J., Diwan, B. A. & Waalkes, M. P. (2010b). Arsenic exposure in utero and nonepidermal proliferative response in adulthood in Tg.AC mice. *International journal of toxicology, 29*, 291-296.

Tokar, E. J., Diwan, S. A. & Waalkes, M. P. (2010c). Arsenic Exposure Transforms Human Epithelial Stem/Progenitor Cells into a Cancer Stem-like Phenotype. *Environmental health perspectives, 118*, 108-115.

Tokar, E. J., Qu, W., Liu, J., Liu, W., Webber, M. M., Phang, J. M. & Waalkes, M. P. (2010d). Arsenic-specific stem cell selection during malignant transformation. *Journal of the National Cancer Institute, 102*, 638-649.

Tseng, C. H., Tai, T. Y., Chong, C. K., Tseng, C. P., Lai, M. S., Lin, B. J., Chiou, H. Y., Hsueh, Y. M., Hsu, K. H. & Chen, C. J. (2000). Long-term arsenic exposure and incidence of non-insulin-dependent diabetes mellitus: A cohort study in arseniasis-hyperendemic villages in Taiwan. *Environmental health perspectives, 108*, 847-851.

Valko, M., Morris, H. & Cronin, M. T. (2005). Metals, toxicity and oxidative stress. *Current medicinal chemistry, 12*, 1161-1208.

Vijayaraghavan, M., Wanibuchi, H., Karim, R., Yamamoto, S., Masuda, C., Nakae, D., Konishi, Y. & Fukushima, S. (2001). Dimethylarsinic acid induces 8-hydroxy-2'-deoxyguanosine formation in the kidney of NCI-Black-Reiter rats. *Cancer letters, 165*, 11-17.

Visvader, J. E. & Lindeman, G. J. (2008). Cancer stem cells in solid tumours: accumulating evidence and unresolved questions. *Nature reviews Cancer, 8*, 755-768.

Waalkes, M. P., Liu, J. & Diwan, B. A. (2007). Transplacental arsenic carcinogenesis in mice. *Toxicology and applied pharmacology, 222*, 271-280.

Waalkes, M. P., Liu, J., Germolec, D. R., Trempus, C. S., Cannon, R. E., Tokar, E. J., Tennant, R. W., Ward, J. M. & Diwan, B. A. (2008). Arsenic exposure in utero exacerbates skin cancer response in adulthood with contemporaneous distortion of tumor stem cell dynamics. *Cancer research, 68*, 8278-8285.

Waalkes, M. P., Liu, J., Ward, J. M. & Diwan, B. A. (2006a). Enhanced urinary bladder and liver carcinogenesis in male CD1 mice exposed to transplacental inorganic arsenic and postnatal diethylstilbestrol or tamoxifen. *Toxicology and applied pharmacology, 215*, 295-305.

Waalkes, M. P., Liu, J., Ward, J. M., Powell, D. A. & Diwan, B. A. (2006b). Urogenital carcinogenesis in female CD1 mice induced by in utero arsenic exposure is exacerbated by postnatal diethylstilbestrol treatment. *Cancer research, 66,* 1337-1345.

Waalkes, M. P., Ward, J. M. & Diwan, B. A. (2004). Induction of tumors of the liver, lung, ovary and adrenal in adult mice after brief maternal gestational exposure to inorganic arsenic: promotional effects of postnatal phorbol ester exposure on hepatic and pulmonary, but not dermal cancers. *Carcinogenesis, 25,* 133-141.

Waalkes, M. P., Ward, J. M., Liu, J. & Diwan, B. A. (2003). Transplacental carcinogenicity of inorganic arsenic in the drinking water: induction of hepatic, ovarian, pulmonary, and adrenal tumors in mice. *Toxicology and applied pharmacology, 186,* 7-17.

Wanibuchi, H., Hori, T., Meenakshi, V., Ichihara, T., Yamamoto, S., Yano, Y., Otani, S., Nakae, D., Konishi, Y. & Fukushima, S. (1997). Promotion of rat hepatocarcinogenesis by dimethylarsinic acid: association with elevated ornithine decarboxylase activity and formation of 8-hydroxydeoxyguanosine in the liver. *Jpn J Cancer Res, 88,* 1149-1154.

Wanibuchi, H., Salim, E. I., Kinoshita, A., Shen, J., Wei, M., Morimura, K., Yoshida, K., Kuroda, K., Endo, G. & Fukushima, S. (2004). Understanding arsenic carcinogenicity by the use of animal models. *Toxicology and applied pharmacology, 198,* 366-376.

Wicha, M. S., Liu, S. & Dontu, G. (2006). Cancer stem cells: an old idea--a paradigm shift. *Cancer research, 66,* 1883-1890; discussion 1895-1886.

Xia, J., Li, Y. J., Yang, Q. L., Mei, C. Z., Chen, Z. W., Bao, B., Ahmad, A., Miele, L., Sarkar, F. H. & Wang, Z. W. (2012). Arsenic Trioxide Inhibits Cell Growth and Induces Apoptosis through Inactivation of Notch Signaling Pathway in Breast Cancer. *Int J Mol Sci, 13,* 9627-9641.

Yamanaka, K., Hasegawa, A., Sawamura, R. & Okada, S. (1991). Cellular response to oxidative damage in lung induced by the administration of dimethylarsinic acid, a major metabolite of inorganic arsenics, in mice. *Toxicology and applied pharmacology, 108,* 205-213.

Yilmaz, O. H., Valdez, R., Theisen, B. K., Guo, W., Ferguson, D. O., Wu, H. & Morrison, S. J. (2006). Pten dependence distinguishes haematopoietic stem cells from leukaemia-initiating cells. *Nature, 441,* 475-482.

Yoda, A., Toyoshima, K., Watanabe, Y., Onishi, N., Hazaka, Y., Tsukuda, Y., Tsukada, J., Kondo, T., Tanaka, Y. & Minami, Y. (2008). Arsenic trioxide augments Chk2/p53-mediated apoptosis by inhibiting oncogenic Wip1 phosphatase. *The Journal of biological chemistry, 283,* 18969-18979.

Yorifuji, T., Tsuda, T. & Grandjean, P. (2010). Unusual cancer excess after neonatal arsenic exposure from contaminated milk powder. *Journal of the National Cancer Institute*, *102*, 360-361.

Yuan, Y., Marshall, G., Ferreccio, C., Steinmaus, C., Liaw, J., Bates, M. & Smith, A. H. (2010). Kidney cancer mortality: fifty-year latency patterns related to arsenic exposure. *Epidemiology*, *21*, 103-108.

Zhang, Z., Wang, X., Cheng, S. P., Sun, L. J., Son, Y. O., Yao, H., Li, W. Q., Budhraja, A., Li, L., Shelton, B. J., et al. (2011). Reactive oxygen species mediate arsenic induced cell transformation and tumorigenesis through Wnt/beta-catenin pathway in human colorectal adenocarcinoma DLD1 cells. *Toxicology and applied pharmacology*, *256*, 114-121.

In: Arsenic   ISBN: 978-1-63321-054-7
Editor: Marissa Jane Olson   © 2014 Nova Science Publishers, Inc.

Chapter 3

ANALYTICAL DETECTION OF ARSENIC SPECIES IN ENVIRONMENTAL AND BIOLOGICAL MATERIALS

Akira Namera[1] and Akito Takeuchi[2]

[1]Department of Forensic Medicine, Institute of Biomedical and Health Sciences, Hiroshima University, Hiroshima, Japan
[2]Osaka Occupational Health Service Center,
Japan Industrial Safety and Health Association, Osaka, Japan

ABSTRACT

Groundwater pollution by inorganic arsenicals is common in South Asia, Southeast Asia, and South America. An arsenic-containing drug (Trisenox®) is used for the treatment of acute promyelocytic leukemia. The oxidation state (valence) of an arsenic species changes upon absorption by a living entity, and these species are metabolized by methylation/alkylation. Various chemical forms of arsenic are therefore widely distributed in environmental and biological materials. A simple and rapid method for the routine monitoring and chemical identification of arsenicals, based on their environmental and clinical properties, is needed because the toxicological and physiological behaviors of arsenicals depend not only on their valences but also on their methylation/alkylation states. Numerous methods have been developed for identifying arsenic compounds in environmental and biological materials. In this chapter, analytical methods such as colorimetric and spectrometric determination and chromatographic separation are summarized, and the advantages and disadvantages are pointed out, to enable accurate routine analysis.

INTRODUCTION

Arsenic exists in various forms and is widely distributed in environmental and biological materials (Figure 1). Recently, drinking water pollution by inorganic arsenic compounds has become a serious problem, causing increased risks to human health, or deaths, in South Asia, Southeast Asia, and South America [1–4]. Vomiting, diarrhea, and abdominal pain are well-known acute symptoms of ingestion of arsenicals; exfoliative dermatitis, neuropathy, and renal failure are chronic symptoms [5–7]. Although its use has been restricted globally, arsenic is still used industrially, e.g., in light-emitting diodes or transistors, because it is one of the most important elements. Moreover, Trisenox®, an arsenic-containing drug, is used for the treatment of acute promyelocytic leukemia [8, 9]. The toxicity of an arsenical depends on its speciation [oxidation state (valence) and methylation/alkylation state]; trivalent inorganic arsenic species are the most toxic to humans and other animals. Arsenic trioxide is the most toxic compound, and has been used for homicide and/or suicide.

After trivalent arsenic is absorbed into the human body, it is metabolized to organic compounds such as monomethylarsenic acid (MMA) and dimethylarsenic acid (DMA) in the liver and excreted in urine (Figure 2) [10–12]. Non-toxic organic arsenic compounds such as arsenosugars and arsenobetaine are also dietary components; therefore, it is necessary to investigate arsenic speciation to evaluate the risk from arsenic. A simple and rapid method is required for the routine monitoring and chemical identification of arsenic compounds, based on their environmental and clinical properties. Arsenite [arsenious acid, As(III)], arsenate [arsenic acid, As(V)], MMA, and DMA are suitable as the target compounds to assess exposure to arsenic. Numerous methods have been developed and reviewed for identifying these arsenic compounds in environmental and biological materials [13–16]; typical strategy for analysis of arsenicals is summarized in Figure 3. Their methods used for separation, detection and extraction are reviewed in this chapter.

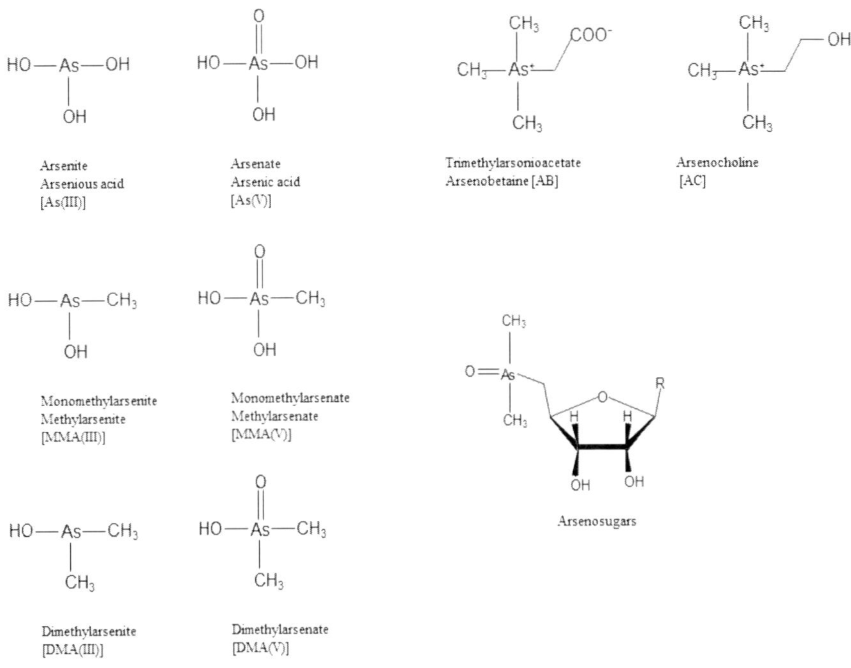

Figure 1. Chemical structure of arsenicals.

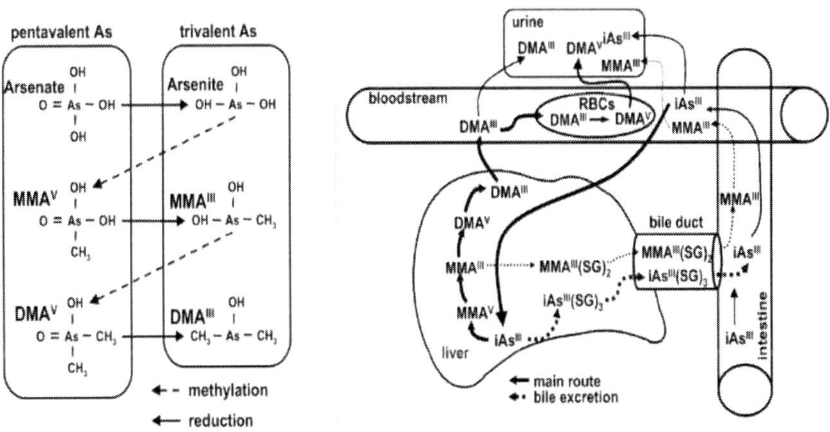

Figure 2. Reduction and methylation reactions of arsenicals, and metabolic and excreted cycle of arsenicals in body. Permission with reference 12.

SEPARATION METHODS

As described in Introduction, arsenic exists in various forms in environmental and biological materials. In order to obtain an accurate result, the complete separation of each compound is required even if for the routine analysis. Gas chromatography, liquid chromatography, and electrophoresis are usually used for separation of various arsenicals although the detecting method to be used also is restricted.

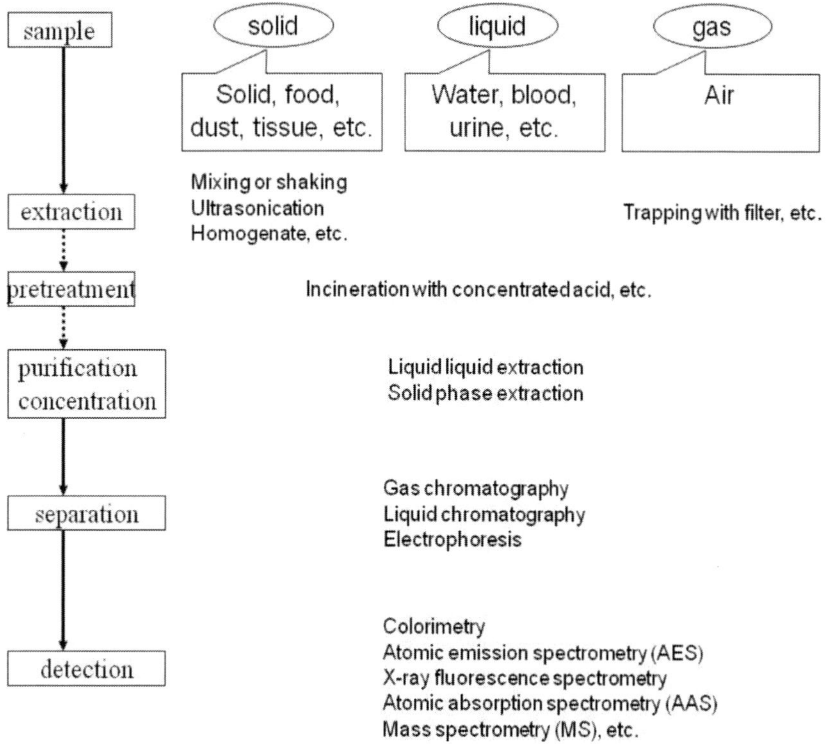

Figure 3. Strategies for arsenic speciation in different matrixes.

Gas Chromatography

The direct detection of arsenicals using gas chromatography (GC) is difficult, because arsenicals have low vaporization and low chromatographic resolution. Arsenicals are usually derivatized to improve the volatility and GC resolution. Trimethylsilyl compounds, methyl thioglycolate (TGM), and dithiol are used as derivatizing agents [17–22].

Each derivatization method has advantages and disadvantages. Trimethylsilylation of an analyte is usually used to improve the characteristics of analyte for GC. N,O-bis(trimethylsilyl)trifluoroacetamide has been used for the derivatization of As(III), As(V), and DMA(V) [17, 18]. The rapid hydrolysis of these derivatives precludes their use in the routine analysis of arsenicals. TGM is one of the most popular derivatizing reagents; it can be used to derivatize inorganic and organic arsenicals simultaneously [19–22]. However, this reagent cannot be used to distinguish trivalent and tetravalent arsenic compounds; the same derivatives are formed from trivalent and tetravalent arsenic by this reagent because of its high reducing power [20]. Moreover, the sensitivities for inorganic arsenicals are lower than those for organic arsenicals, because inorganic arsenic derivatives can decompose in the GC injection port [21]. 2,3-Dithiopropanol (British anti-Lewisite, BAL), which is well known as an antidote to arsenic poisoning, is also used as a derivatizing agent for arsenic analysis [23, 24]. BAL reacts with trivalent arsenic at room temperature. Speciation analysis of inorganic and organic arsenicals in urine, using a combination of reduction processes, has been reported [25, 26]. When arsenic is analyzed using GC, it is necessary to choose a derivatization method.

Liquid Chromatography

High-performance liquid chromatography (HPLC) is a powerful tool for the analysis of non-volatile compounds. However, complete separation is required for arsenic analysis, because similar analogs may be present in samples. Many strategies have been developed for achieving complete separation of such analogs. One is mobile-phase selection, and the other is column selection. Arsenicals are hydrophilic; inorganic arsenicals, in particular, are highly water soluble and have a low reverse-phase (RP) affinity. An ion-pair reagent is added to the mobile phase to increase the affinity for the RP column [27–29]. At neutral pH, As(V) (pKa_1 = 2.3), MMA(V) (pKa_1 = 3.6), and DMA(V) (pKa = 6.2) are present as anions, arsenocholine and tetramethylarsonium ions as cations, arsenobetaine as a zwitterion, and As(III) (pKa_1 = 9.3) as an uncharged species. Cation- or anion-exchange moieties are therefore selected to improve the affinity for the analytical column; anion-exchange [30–35] and cation-exchange columns [31–35] are commonly used for the separation of ionic arsenic species (Figure 4).

Many reports have been published on the separation of As(III), As(V), MMA(V), and DMA(V), which are negatively charged, using cation-exchange columns; however, zwitterionic compounds such as arsenobetaine showed poor affinity for such columns, or high retention. However, inorganic and organic arsenicals were coeluted and speciation separation was not successful (Figure 4b), because the concentration of organic arsenicals was much higher than that of inorganic arsenicals. Although it is also possible to calculate the difference between the amounts of organic and inorganic arsenicals, the reproducibility may be poor.

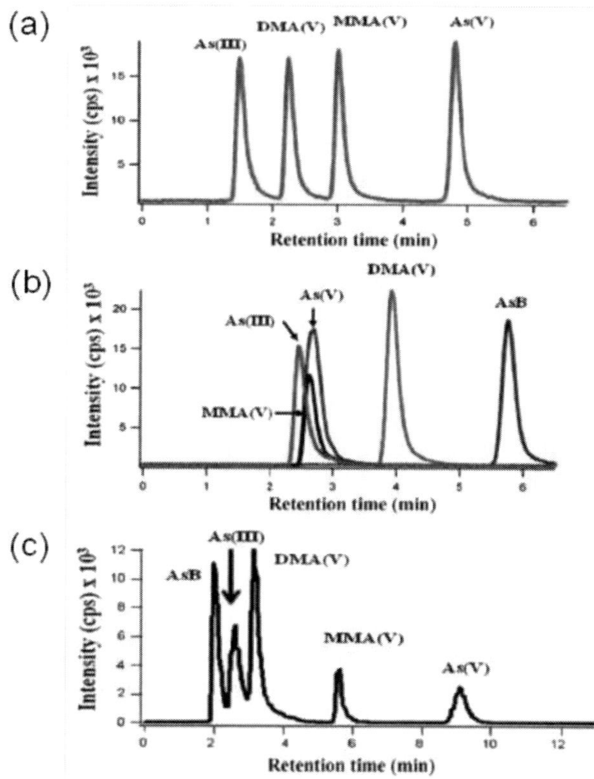

Figure 4. Chromatograms of tested arsenicals (5 µg/L) in spiked water obtained by HPLC-ICP/MS. (a) Ion pair separation. (b) cation exchange separation. (c) anion exchange separation. Permission with reference 35.

Many reports have been published on the use of anion-exchange columns to improve selectivity. As(III), As(V), MMA(V), and DMA(V) are completely separated on such columns (Figure 4c). Seafood consumption increases the amount of organic arsenic in urine, but good separation was obtained even for these samples.

Capillary Electrophoresis

Although HPLC provides ultra-sensitivity and multi-element separation, it is expensive and has limited separation efficiency. Recently, there has been increasing interest in the use of capillary electrophoresis (CE) for the separation and determination of arsenic species, because CE has the advantages of short analysis time, high separation efficiency, and low operating costs. In recent years, CE has been used successfully to separate inorganic, organic, and methylated arsenic compounds simultaneously [36–39].

DETECTION METHODS

Many detection methods are applied for the determination of arsenicals in various samples [13, 15] and these detection methods are compared for the determination of arsenicals in some studies [40–44]. Although colorimetric determination is simple and suitable for on-site detection, this method is lack of selectivity and sensitivity and the confirmed determination is needed for identification and speciation of arsenicals. AES and AAS which use for the routine analysis are sensitive, however these detection methods are lack of the selectivity because the speciation is not confirmed by using both methods. Recently mass spectrometry has been used for the routinely analysis of arsenicals. The detection mechanism and remark of each method are simply reviewed in the following section.

Colorimetric Detection

Colorimetric methods are useful for the determination of arsenic in samples, and screening can be performed by non-specialists, because special techniques and instruments are not required. Reinsch developed a visible detection method for arsenic more than 100 years ago [45]. In this method, a thin copper arsenide layer is formed on a copper surface using highly concentrated hydrogen chloride. If arsenic is present in the sample, the copper surface changes color, and this can be observed by the naked eye.

Marsh and Gutzeit developed other methods. In their methods, arsine is generated using zinc and concentrated acid. In the Marsh test, the generated arsine is decomposed by heating, and converted to arsenic metal [46]. The arsenic metal is adsorbed on a glass surface, forming a shiny black powder (an arsenic mirror). In the Gutzeit method [47, 48], the generated arsine is reacted with mercury(II) bromide, and the color of the test paper changes from yellow to brown. Arsine itself is colorless and odorless, but after oxidation by air it smells slightly of garlic. The above methods are highly sensitive, but the selectivity is low because mercury and antimony can also undergo the same reactions. Other colorimetric methods, e.g., using diethyldithiocarbamate and molybdenum blue, have been reported for arsenic determination in environmental samples. Diethyldithiocarbamate quantitatively forms a water-soluble, red-purple complex with arsine in pyridine solution [49]. The quantitative limit for arsenic is 1.0 μg. Molybdenum ammonium forms a heteropoly acid and blue complex with arsine [50]. The detection limit of this method is 20 μg As/L.

Atomic Emission Spectrometry

In atomic emission spectrometry (AES), an emission spectrum is produced when a species returns from an excited state to the ground state. Metals or chemical compounds are excited by a flame, plasma, arc, or spark. Many elements can be determined simultaneously using this method. The intensity of the spectrum depends on the amount of the species in the sample; therefore, arsenic can be quantified using this method. AES is a highly sensitive detection method, but the arsenic valences are indistinguishable. The arsenic species therefore need to be separated using a selective extraction or chromatographic technique before AES [51]. The detection limits in water using anion-exchange chromatography coupled with AES are 0.35, 0.92, 0.38, and 2.13 ng for As(III), As(V), MMA(V), and DMA(V), respectively[52].

X-Ray Fluorescence Spectrometry

X-ray fluorescence analysis (XRF) is also a simple determination method [53–55]. In XRF, the intensity and wavelength of the spectral line from each element are used instead of the emission spectrum. Samples are directly analyzed using XRF, and tedious sample preparation is not required for arsenic determination. However, simple preconcentration is required because the sensitivity is lower than that with AES. The limits of detection in water and urine are 0.02 and 0.05 mg/L, respectively.

Atomic Absorption Spectrometry

In atomic absorption spectrometry (AAS), the target element is analyzed by examination of the adsorption wavelength after atomization of the sample in a flame or electrothermal (graphite tube) atomizer. In general, each wavelength corresponds to only one element, and the intensity depends on the amount of element in the sample. Elements can be identified and quantified using this method, but elements in organic chemicals are required to incinerate using high concentrated acid or reduce using sodium tetrahydroboate, and the hydride generation procedure is often required for arsenic analysis.

The selectivity of this method is extremely high. Elements can only be measured individually. The sensitivity of this method is lower than those of AES and mass spectrometry (MS). The limits of detection for arsenic in water and urine are 0.46 and 2.0 µg/L, respectively, under the optimized conditions [56, 57].

Mass Spectrometry

In MS, the sample is ionized using an ion source, and a characteristic spectrum of the masses of the atoms or molecules in the sample is obtained. The spectrum is based on measurements of the mass-to-charge ratios of the charged molecules or molecular fragments produced by ionization. The atoms or molecules in the sample can be identified by correlating known masses with those in the spectrum or using characteristic fragmentation patterns. The intensity of the spectrum depends on the amount of the component in the sample; therefore, arsenic can be quantified using this method.

The ionization method is important in determining the sample types that can be analyzed by MS. Several ionization methods are known to produce atomic or molecular ions. Electron ionization (EI) and chemical ionization are usually used for biological samples. Although organic arsenicals are sufficiently ionized by EI, the ionization efficiency of inorganic arsenic by EI is low. Inductively coupled plasma (ICP) is therefore widely used for the ionization of inorganic arsenic. ICP-MS is suitable and sensitive for the determination of metallic elements, but the selectivity is low because only the mass-to-charge ratio is monitored [58–60], so species with the same mass-to-charge ratios are indistinguishable. Complete separation of the species being analyzed is therefore required, to increase the selectivity.

Chemical separation analysis using HPLC coupled with ICP-MS (HPLC/ICP-MS) is currently popular in toxicological analyses, because the toxicological and physiological behaviors of arsenicals depend not only on their oxidation states (valences) but also on their methylation/alkylation states. The limits of detection for As(III), As(V), MMA, DMA in urine are 0.3, 0.2, 0.2, and 0.2 μg/L, respectively, by HPLC/ICP-MS in the anion mode [33]. However, HPLC/ICP-MS systems are not available in all laboratories, because of the high acquisition, maintenance, and running costs of such equipment. Although the valences of arsenic compounds can be identified using HPLC/ICP-MS, the peak of trivalent arsenite [As(III)] sometimes overlaps those of DMA(V), arsenobetaine, and an unknown arsenic compound [18–20, 33, 35]. As a result, the obtained concentrations of As(III) were higher than the true concentrations.

EXTRACTION

Inorganic and methylated arsenicals are highly hydrophilic and it is difficult to extract them from samples using classical liquid–liquid extraction. Arsenicals can be extracted from liquid samples after the formation of ion-associated complexes with diethyldithiocarbamate (DDC) or molybdenum acid [61, 62]. The selective extraction of As(III), As(V), and MMA is achieved using a combination of a chloride or DDC system and a pyrogallol/ tetraphenylarsonium chloride system. MMA is also separated using an iodide system. Another strategy involves arsenical extraction after derivatization with a dithiol such as BAL [23, 25, 26]. This method is cost effective and simple. DMA is not derivatized with BAL and is neither extracted using this method. Although other advanced techniques for the extraction of arsenicals from water samples have also been reported [63–65], there are few methods for arsenical extraction from biological samples, because endogenous substances may interfere with liquid–liquid extraction.

These problems are overcome using cartridges containing packing materials that have been found to be effective for extraction of arsenicals: C18, silica, and alumina cartridges. No retention of arsenic species was observed, except with a quaternary methylamine and alumina cartridges. Although alumina cartridges are suitable for the extraction of arsenicals, and MMA and DMA are eluted almost completely, it is very difficult to elute inorganic arsenic from these cartridges using a mixture of NaOH and NaCl. For the elution of As(V) and As(III), 2.0 M hydrofluoric acid is necessary. Because of the differences between their elution behaviors, MMA and DMA can be separated from inorganic As(III) and As(V) using an alumina cartridge and selective elution [66, 67].

ARSENIC STABILITY

Although many accurate and reliable methods have been developed for quantitative and qualitative analyses of arsenicals in environmental and biological materials, speciation changes can occur during arsenical storage and extraction for analysis. Many studies have been undertaken to determine the best conditions for avoiding speciation changes, but appropriate conditions for long-term storage have not yet been identified; the stabilities of arsenicals remain problematic.

In the spiked real water samples with low (0.5–20 µg/L) concentrations of As(III) and As(V), As(V) was reduced to As(III) within a few days [68]. As(III) solutions were unstable and underwent complete conversion to As(III) within 4 days [69]. The oxidation of As(III) or reduction of As(V) in this mixed sample was pH dependent, and occurred within a few hours under acidic conditions or within days at pH 7 [70].

Changes in speciation or the total amount of arsenic during storage of biological specimens have also been reported. As(III), As(V), MMA(V), DMA(V), and arsenobetaine in urine were stable for up to 2 months, without any additives, when the urine samples were stored at 4 and −20 °C [71]. The arsenicals in freshly collected urine samples (pH = 5.5–7.0) and freeze-dried urine (National Institute of Standards and Technology Standard Reference Material 2670, pH 4.4) both remained unchanged for up to 6 months when stored at −20 °C. In an aqueous solution mixed with 10 µg/L of As(III), As(V),

MMA, and DMA standards stored at 4 °C, As(III) and As(V) were stable for only up to 4 weeks, but MMA and DMA were stable for up to 4.5 months [72].

CONCLUSION

Highly selective and sensitive methods have been developed for the determination of trace amounts of arsenicals in environmental and biological samples. Speciation analysis is also required, because arsenicals of different toxicities may be present in samples. Accurate analysis is required for the identification of each arsenical species, but complete separation and identification is impossible, because of speciation changes under storage conditions and in matrixes. On-site testing is important for the detection of arsenicals in environmental and biological samples, because their stabilities in samples are still controversial. It is impossible to perform rapid speciation analysis at all research institutes. Even if highly sensitive instruments can be used for arsenical determination, analysis of standardized samples (standard reference materials) is indispensable.

REFERENCES

[1] Tondel, M., Rahman, M., Magnuson, A., Chowdhury, I. A., Faruquee, M. H. & Ahmad, S. A. (1999). The relationship of arsenic levels in drinking water and the prevalence rate of skin lesions in Bangladesh. *Environmental Health Perspectives, 107,* 727–729.

[2] Berg, M., Tran, H. C., Nguyen, T. C., Pham, H. V., Schertenleib, R. & Giger, W. (2001). Arsenic contamination of groundwater and drinking water in Vietnam: A human health threat. *Environmental Science and Technology, 35,* 2621–2626.

[3] McClintock, T. R., Chen, Y., Bundschuh, J., Oliver, J. T., Navoni, J., Olmos, V., Lepori, E. V., Ahsan, H. & Parvez, F. (2012). Arsenic exposure in Latin America: Biomarkers, risk assessments and related health effects. *Science of the Total Environ*ment, *429,* 76–91.

[4] Wright, J. A., Yang, H., Rivett, U. & Gundry, S. W. (2012). Public perception of drinking water safety in South Africa 2002–2009: a repeated cross-sectional study. *BMC Public Health, 12,* 556.

[5]   Basu, A., Saha, D., Saha, R., Ghosh, T. & Saha, B. (2012). A review on sources, toxicity and remediation technologies for removing arsenic from drinking water. *Research on Chemical Intermediates*, *40*, 447–485.

[6]   Jomova, K., Jenisova, Z., Feszterova, M., Baros, S., Liska, J., Hudecova, D., Rhodes, C. J. & Valko, M. (2011). Arsenic: toxicity, oxidative stress and human disease. *Journal of Applied Toxicology*, *31*, 95–107.

[7]   Tournel, G., Houssaye, C., Humber, L., Dhorn, C., Gnemmi, V., Bécart-Robert, A., Nisse, P., Hédouin, V., Gosset, D. & Lhermitte, M. (2011). Acute arsenic poisoning: clinical, toxicological, histopathological, and forensic features. *Journal of Forensic Science*, *56*, S275–S279.

[8]   Douer, D., Hu, W., Giralt, S., Lill, M. & DiPersio, J. (2003). Arsenic trioxide (Trisenox®) therapy for acute promyelocytic leukemia in the setting of hematopoietic stem cell transplantation. *Oncologist*, *8*, 132–140.

[9]   Held, L. A., Rizzieri, D., Long, G. D., Gockerman, J. P., Diehl, L. F., de Castro, C. M., Moore, J. O., Horwitz, M. E., Chao, N. J. & Gasparetto, C. (2013). A phase I study of arsenic trioxide (Trisenox), ascorbic acid, and Bortezomib (Velcade) combination therapy in patients with relapsed/refractory multiple myeloma. *Cancer Investigation*, *31*, 172–176.

[10]   Aposhian, H. V., Zakharyan, R. A., Avram, M. D., Sampayo-Reyes, A. & Wollenberg, M. L. (2004). A review of the enzymology of arsenic metabolism and a new potential role of hydrogen peroxide in the detoxication of the trivalent arsenic species. *Toxicology and Applied Pharmacology*, *198*, 327–335.

[11]   Vahter, M. (2002). Mechanisms of arsenic biotransformation. *Toxicology*, *181–182*, 211–217.

[12]   Suzuki, K. T., Mandal, B. K. & Ogra, Y. (2002). Speciation of arsenic in body fluids. *Talanta*, *58*, 111-119.

[13]   Gong, Z., Lu, X., Ma, M., Watt, C. & Le, X. C. (2002). Arsenic speciation analysis. *Talanta*, *58*, 77–96.

[14]   Francesconi, K. & Kuehnelt, D. (2004). Determination of arsenic species: A critical review of methods and applications, 2000–2003. *Analyst*, *129*, 373–395.

[15]   Hung, D. Q., Nekrassova, O. & Compton, R. G. (2004). Analytical methods for inorganic arsenic in water: a review. *Talanta*, *64*, 269–277.

[16]   Hsu, K. C., Sun, C. C. & Huang, Y. L. (2011). Arsenic speciation in biomedical sciences: recent advances and applications. *Kaohsiung Journal of Medical Sciences*, *27*, 382-389.

[17] Henry, F. T. & Thorpe, T. M. (1978). Gas chromatography of the trimethylsilyl derivatives of arsenic, arsenious, and dimethylarsinic acids. *Journal of Chromatography A, 166*, 577–586.

[18] Butts, W. C. & Rainey Jr., W. T. (1971). Gas chromatography and mass spectrometry of the trimethylsilyl derivatives of inorganic anions. *Analytical Chemistry, 43*, 538–542.

[19] Dix, K., Cappon, C. J. & Toribara, T. Y. (1987). Arsenic speciation by capillary gas-liquid chromatography. *Journal of Chromatographic Science, 25*, 164–169.

[20] Claussen, F. A. (1997). Arsenic speciation of aqueous environmental samples by derivatization with thioglycolic acid methylester and capillary gas-liquid chromatography-mass spectrometry. *Journal of Chromatographic Science, 35*, 568–572.

[21] Szostek, B. & Aldstadt, J. H. (1998). Determination of organoarsenicals in the environment by solid-phase microextraction–gas chromatography–mass spectrometry. *Journal of Chromatography A, 807*, 253–263.

[22] Campillo, N., Peñalver, R., Viñas, P., López-García, I. & Hernández-Córdoba, M. (2008). Speciation of arsenic using capillary gas chromatography with atomic emission detection. *Talanta, 77*, 793–799.

[23] Fukui, S., Hirayama, T., Nohara, M. & Sakagami, Y. (1983). Determination of arsenite, arsenate and monomethylarsonic acid in aqueous samples by gas chromatography of their 2,3-dimercaptopropanol (bal) complexes. *Talanta, 30*, 89–93.

[24] Siu, K. W. M., Roberts, S. Y. & Berman, S. S. (1984). Derivatization and determination of arsenic in marine samples by gas chromatography with electron capture detection. *Chromatographia, 19*, 398–400.

[25] Takeuchi, A., Namera, A., Kawasumi, Y., Imanaka, T., Sakui, N., Ota, H., Endo, Y., Sumino, K. & Endo, G. (2012). Development of an analytical method for the determination of arsenic in urine by gas chromatography-mass spectrometry for biological monitoring of exposure to inorganic arsenic. *Journal of Occupational Health, 54*, 434–440.

[26] Namera, A., Takeuchi, A., Saito, T., Miyazaki, S., Oikawa, H., Saruwatari, T. & Nagao, M. (2012). Sequential extraction of inorganic arsenic compounds and methyl arsenate in human urine using mixed-mode monolithic silica spin column coupled with gas chromatography-mass spectrometry. *Journal of Separation Science, 35*, 2506–2513.

[27] Larsen, E. H. & Hansen, S. H. (1992). Separation of arsenic species by ion-pair and ion exchange high performance liquid chromatography. *Mikrochimica Acta, 109,* 47–51.

[28] Le, X. C., Ma, M. & Wong, N. A. (1996). Speciation of arsenic compounds using high-performance liquid chromatography at elevated temperature and selective hydride generation atomic fluorescence detection. *Analytical Chemistry, 68,* 4501–4506.

[29] Do, B., Robinet, S., Pradeau, D. & Guyon, F. (2001). Speciation of arsenic and selenium compounds by ion-pair reversed-phase chromatography with electrothermic atomic absorption spectrometry: Application of experimental design for chromatographic optimization. *Journal of Chromatography A, 918,* 87–98.

[30] Day, J. A., Montes-Bayón, M., Vonderheide, A. P. & Caruso, J. A. (2002). A study of method robustness for arsenic speciation in drinking water samples by anion exchange HPLC-ICP-MS. *Analytical and Bioanalytical Chemistry, 373,* 664–668.

[31] Falk, K. & Emons, H. (2000). Speciation of arsenic compounds by ion-exchange HPLC-ICP-MS with different nebulizers. *Journal of Analytical Atomic Spectrometry, 15,* 643–649.

[32] Kirby, J., Maher, W., Ellwood, M. & Krikowa, F. (2004). Arsenic species determination in biological tissues by HPLC–ICP–MS and HPLC–HG–ICP–MS. *Australian Journal of Chemistry, 57,* 957–966.

[33] Suzuki, Y., Shimoda, Y., Endo, Y., Hata, A., Kenzo, Y., Endo, G. (2009). Rapid and effective speciation analysis of arsenic compounds in human urine using anion-exchange columns in HPLC-ICP-MS. *Journal of Occupational Health, 51,* 380–385.

[34] Ciardullo, S., Aureli, F., Raggi, A. & Cubadda, F. (2010). Arsenic speciation in freshwater fish: Focus on extraction and mass balance. *Talanta, 81,* 213–221.

[35] Chen, L. W. L., Lu, X. & Le, X. C. (2010). Complementary chromatography separation combined with hydride generation–inductively coupled plasma mass spectrometry for arsenic speciation in human urine. *Analytica Chimica Acta, 675,* 71–75.

[36] Yin, X. B., Yan, X. P., Jiang, Y. & He, X. W. (2002). On-line coupling of capillary electrophoresis to hydride generation atomic fluorescence spectrometry for arsenic speciation analysis. *Analytical Chemistry, 74,* 3720–3725.

[37] Gil, E. P., Ostapczuk, P. & Emons, H. (1999). Determination of arsenic species by field amplified injection capillary electrophoresis after

modification of the sample solution with methanol. *Analytica Chimica Acta, 389*, 9–19.

[38] Sun, B., Macka, M. & Haddad, P. R. (2002). Separation of organic and inorganic arsenic species by capillary electrophoresis using direct spectrophotometric detection. *Electrophoresis, 23*, 2430–2438.

[39] Sun, B., Macka, M. & Haddad. P. R. (2004). Speciation of arsenic and selenium by capillary electrophoresis. *Journal of Chromatography A, 1039*, 201–208.

[40] Kumaresan, M. & Riyazuddin, P. (2001). Overview of speciation chemistry of arsenic. *Current Science, 80*, 837-846.

[41] Simeonsson, J. B., Elwood, S. A., Ezer, M., Pacquette, H. L., Swart, D. J., Beach, H. D. & Thomas, D. J. (2002). Development of ultratrace laser spectrometry techniques for measurements of arsenic. *Talanta, 58*, 189-199.

[42] Melamed, D. (2005). Monitoring arsenic in the environment: a review of science and technologies with the potential for field measurements. *Analytica Chinica Acta, 532*, 1-13.

[43] Akter, K. F., Chen, Z., Smith, L., Davey, D. & Naidu, R. (2005). Speciation of arsenic in ground water samples: a comparative study of CE-UV, HG-AAS and LC-ICP-MS. *Talanta, 68*, 406-415.

[44] Lindberg, A. L., Goessler, W., Grander, M., Nermell, B. & Vahter, M. (2007). Evaluation of the three commonly used analytical methods for determination of inorganic arsenic and its metabolites in urine. *Toxicology Letters, 168*, 310-318.

[45] Reinsch, H. (1841). On the action of metallic copper on solutions of certain metals, particularly with reference to the detection of arsenic. *Philosophical Magazine, 19*, 480–483.

[46] Marsh, J. (1836). Account of a method of separating small quantities of arsenic from substances with which it may be mixed. *Edinburgh New Philosophical Journal, 21*, 229–236.

[47] Jacobs, M. B. & Nagler, J. (1942). Colorimetric microdetermination of arsenic. *Industrial and Engineering Chemistry, 14*, 442–444.

[48] Rahman, M. M., Seike, Y. & Okumura, M. (2006). Concentrations of arsenic in brackish lake water: application of tristimulus colorimetric determination. *Analytical Sciences, 22*, 475–478.

[49] Lakso, J. U., Rose, L. J., Peoples, S. A. & Shirachi, D. Y. (1979). A colorimetric method for the determination of arsenite, arsenate, monomethylarsonic acid, and dimethylarsinic acid in biological and

environmental samples. *Journal of Agricultural and Food Chemistry*, *27*, 1229–1233.

[50] Lenoble, V., Deluchat, V., Serpaud, B. & Bollinger, J. (2003). Arsenite oxidation and arsenate determination by the molybdene blue method. *Talanta*, *61*, 267–276.

[51] Terlecka, E. (2005). Arsenic speciation analysis in water sample: A review of the hyphenated techniques. *Environmental Monitoring and Assessment*, *107*, 259–284.

[52] Rauret, G., Rubio, R. & Padro, A. (1991). Arsenic speciation using HPLC hydride generation ICP AES with gas liquid separator. *Fresenius Journal of Analytical Chemistry*, *340*, 157–160.

[53] Sbarato, V. M. & Sanchez, H. J. (2001). Analysis of arsenic pollution in groundwater aquifers by X-ray fluorescence. *Application of Radioactive Isotopes*, *54*, 737–740.

[54] Peraniemi, S. & Ahlgren, M. (1995). Optimized arsenic, selenium and mercury determinations in aqueous solutions by energy dispersive x-ray fluorescence after preconcentration onto zirconium-loaded activated charcoal. *Analytica Chimica Acta*, *302*, 89–95.

[55] Marco P, L. M. & Hernandez-Caraballo, E. A. (2004). Direct analysis of biological samples by total reflection X-ray fluorescence. *Spectrochimica Acta, Part B*, *59*, 1077– 1090.

[56] Piston, M., Silva, J., Perez-Zambra, R., Dol, I. & Knochen, M. (2012). Automated method for the determination of total arsenic and selenium in natural and drinking water by HG-AAS. *Environmental Geochemistry and Health*, *34*, 273–278.

[57] Lehmann, E. L., Fostier, A. H. & Arruda, M. A. Z. (2013). Hydride generation using a metallic atomizer after microwave-assisted extraction for inorganic arsenic speciation in biological samples. *Talanta*, *104*, 187–192.

[58] Komorowicz, I. & Baralkiewicz, D. (2011). Arsenic and its speciation in water samples by high performance liquid chromatography inductively coupled plasma mass spectrometry—Last decade review. *Talanta*, *84*, 247–261.

[59] Rivera-Núñez, Z., Linder, A. M., Chen, B. & Nriagu, J. O. (2011). Low-level determination of six arsenic species in urine by high performance liquid chromatography-inductively coupled plasma-mass spectrometry (HPLC-ICP-MS). *Analytical Methods*, *3*, 1122–1129.

[60] Morton, J. & Leese, E. (2011). Arsenic speciation in clinical samples: urine analysis using fast micro-liquid chromatography ICP-MS. *Analytical and Bioanalytical Chemistry, 399,* 1781–1788.

[61] Suzuki, N., Satoh, K., Shoji, H. & Imura, H. (1986). Liquid-liquid extraction behavior of arsenic(III), arsenic(V), methylarsonate and dimethylarsinate in various systems. *Analytica Chimica Acta, 185,* 239–248

[62] Kanke, M., Kumamaru, T., Sakai, K. & Yamamoto, Y. (1991). Determination of arsenic by graphite furnace atomic absorption spectrometry combined with liquid-liquid extraction of arsenomolybdic acid. *Analytica Chimica Acta, 247,* 13–18.

[63] Tang, A. N., Ding, G. S. & Yan, X. P. (2005). Cloud point extraction for the determination of As(III) in water samples by electrothermal atomic absorption spectrometry. *Talanta, 67,* 942–946.

[64] Jiang, H. M., Hu, B., Chen, B. B. & Xia, L. B. (2009). Hollow fiber liquid phase microextraction combined with electrothermal atomic absorption spectrometry for the speciation of arsenic (III) and arsenic (V) in fresh waters and human hair extracts. *Analytica Chimica Acta, 634,* 15–21.

[65] Rivas, R. E., Ignacio, L. G. & Manuel, H. C. (2009). Speciation of very low amounts of arsenic and antimony in waters using dispersive liquid-liquid microextraction and electrothermal atomic absorption spectrometry. *Spectrochimica Acta, Part B, 64,* 329–333.

[66] Yalçin, S. & Le, X. C. (2001). Speciation of arsenic using solid phase extraction cartridges. *Journal of Environmental Monitoring, 3,* 81–85.

[67] Zhang, L., Morita, Y., Sakuragawa, A. & Isozaki, A. (2007). Inorganic speciation of As(III, V), Se(IV, VI) and Sb(III, V) in natural water with GF-AAS using solid phase extraction technology. *Talanta, 72,* 723–729.

[68] Hall, G. E. M., Pelchat, J. C. & Gauthier, G. (1999). Stability of inorganic arsenic (III) and arsenic (V) in water samples. *Journal of Analytical Atomic Spectrometry, 14,* 205–213.

[69] Feldman, C. (1979). Improvements in the arsine accumulation-helium glow detector procedure for determining traces of arsenic. *Analytical Chemistry, 51,* 664–669.

[70] Cherry, J. A., Shaikh, A. U., Tallman, D. E. & Nicholson, R. V. (1979). Arsenic species as an indicator of redox conditions in groundwater. *Journal of Hydrology, 43,* 373–392.

[71] Feldmann, J., Lai, V. W. M., Cullen, W. R., Ma, M., Lu, X. & Le, X. C. (1999). Sample preparation and storage can change arsenic speciation in human urine. *Clinical Chemistry, 45*, 1988–1997.

[72] Chen, Y. C., Amarasiriwardena, C. J., Hsueh, Y. M. & Christiani, D. C. (2002). Stability of arsenic species and insoluble arsenic in human urine. *Cancer Epidemiology, Biomarkers and Prevention, 11*, 1427–1433.

In: Arsenic
Editor: Marissa Jane Olson

ISBN: 978-1-63321-054-7
© 2014 Nova Science Publishers, Inc.

Chapter 4

INORGANIC ARSENIC IN FOODSTUFFS: HEALTH EFFECTS AND ANALYTICAL METHODS FOR ITS DETERMINATION

I. N. Pasias, A. K. Psoma[†], N. I. Rousis[‡], and N. S. Thomaidis[§]*

National and Kapodistrian University of Athens, Department of Chemistry, Laboratory of Analytical Chemistry, Panepistimiopolis Zografou, Athens, Greece

ABSTRACT

Arsenic is one of the most toxic trace elements and occurs in both inorganic and organic forms, in many foodstuffs. The inorganic arsenic forms (As(III) and As(V)) were deemed group I carcinogen by the International Agency for Research on Cancer (IARC), and therefore the World Health Organization recommended 15 µg of total inorganic arsenic (t-inAs) per body weight as a provisional tolerable weekly intake. This book chapter deals with the risk assessment of the inorganic arsenic content in different foodstuffs and especially for the more sensitive

[*] Equal contribution.
[†] Equal contribution.
[‡] Equal contribution.
[§] Nikolaos S. Thomaidis, Laboratory of Analytical Chemistry, Department of Chemistry, University of Athens Panepistimiopolis Zografou, 15771 Athens, Greece. Tel: +30 210 7274317, Fax: +30 210 7274750. E-mail address: ntho@chem.uoa.gr (N.S. Thomaidis).

population groups, such as children. Furthermore, in this chapter there will be presented all the different developed methods, concerning the determination of the inorganic arsenic forms and there will be an overall comparison concerning the ability of the different methods to determine low contents of the toxic arsenic forms in the different food matrices.

ABBREVIATIONS

As Low As Reasonably Achievable	the ALARA principle
Atomic Absorption Spectrometry	AAS
Atomic Fluorescence Spectrometry	AFS
Benchmark Dose Limit	BDL
Dimethylarsinic acid	DMA V
Directorate General for Health and Consumers	DG SANCO
Dispersive Liquid-Liquid Microextraction	DLLME
Electrothermal Atomic Absorption Spectrometry	ETAAS
European Commission	EC
European Food Safety Authority	EFSA
European Union	EU
European Union Reference Laboratory for Heavy Metals in Feed and Food	EU-RL-HM
Food and Agriculture Organization	FAO
Food Standards Australia New Zealand	FSANZ
Gas Chromatography	GC
High Performance Liquid chromatography coupled with Inductively Coupled Plasma – Mass Spectrometry	HPLC-ICP-MS
Hydride generation atomic absorption spectrometry	HG-AAS
Inductively Coupled Plasma–Mass Spectrometry	ICP–MS
Inorganic arsenic	inAs
Institute for Reference Materials and Measurements	IRMM
International Agency for Research on Cancer	IARC
Joint Expert Committee on Food Additives	JECFA
Limit of Detection	LOD
Limit of Quantification	LOQ
Liquid Chromatography	LC
Mass Spectrometry	MS

Maximum Contaminant Level	MCL
Maximum Contaminant Level Goal	MCLG
Maximum Level	MLs
Monomethylarsonic acid	MMA V
Monomethylarsonous acid	MMA III
National Institute of Standards and Technology	NIST
National Research Council of Canada	CNRC
Panel on Contaminants in the Food Chain	CONTAM Panel
Provisional Tolerable Daily Intake	PTDI
Provisional Tolerable Weekly Intake	PTWI
Solid Phase Extraction	SPE
Tolerable Daily Intake	TDI
US Environmental Protection Agency	US EPA
World Health Organization	WHO
World Trade Organization	WTO

1. INTRODUCTION

There are, mainly, two types of sources of the environmental arsenic; the natural processes and the anthropogenic activities. Weathering of rocks, followed by leaching of metals, leads to the introduction of arsenic into soil and water. Two other important natural processes are the biological activity and the volcanic emissions. Humans are exposed to the toxic arsenic (As) species primarily from food and water (Figure 1). Arsenic is considered one of the most important toxic elements found in the environment, which exists mainly in four oxidation states, +V (arsenate), +III (arsenite), 0 (arsenic) and -III (arsine) [1].

The International Agency for Research on Cancer (IARC) has listed As as a human carcinogen since 1980. The Working Group made the overall evaluation on 'arsenic and inorganic arsenic compounds', based on the combined results of epidemiological studies, carcinogenicity studies in experimental animals and data on the chemical characteristics, metabolism and modes of action of carcinogenicity. As a result of this evaluation As and inorganic As compounds are considered carcinogenic to humans (Group 1), dimethylarsinic acid and monomethylarsonic acid as possibly carcinogenic to humans (Group 2B) and arsenobetaine and other organic arsenic compounds not metabolized in humans, are not classifiable as to their carcinogenicity to humans (Group 3) [2].

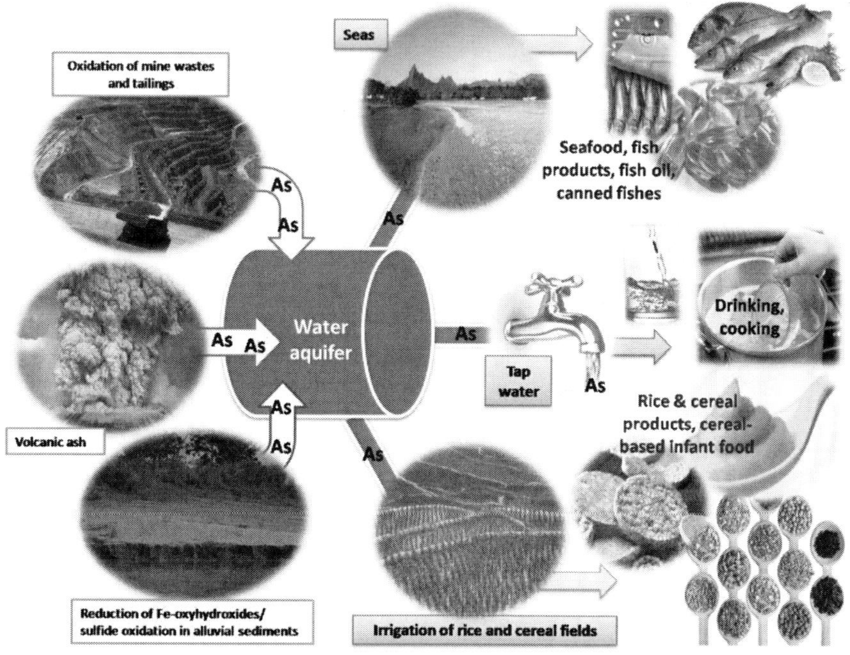

Figure 1. Pathways of human exposure to inorganic As through contamination of water aquifer and food chain.

Toxicity studies that have been done in animals (e.g., mice, rats and beagles) at low and high exposure levels cannot fully provide a suitable basis for risk characterization for humans, because the arsenic metabolism and toxic-kinetics of the experimental animal models differ considerably from humans. Although, there are a few indications that human metabolism can, to a certain degree, adapt to As exposure, there are no studies indicating any beneficial effects of As exposure to humans as is the case with short lived livestock. [3-5].

2. TOXICITY OF INORGANIC ARSENIC

The toxicity of arsenic to organisms depends on its concentration and speciation and in general the organic form of arsenic is less toxic than the inorganic form, arsenite is more toxic than arsenate and the human metabolites of arsenic are more toxic than the parent compounds. Recent findings show the following order of acute toxicity: MMA(III) > As(III) > As(V) > DMA(V) > MMA(V), where MMA(III) is the monomethylarsonous acid, MMA(V) is the monomethylarsonic acid and DMA(V) is the dimethylarsinic acid (Figure 2). However, MMA(III) has not been detected in food or water and therefore, As(III) is considered the most toxic form of arsenic.

In general, As(V) is more readily absorbed than As(III), and inorganic arsenic more easily than organic arsenic. The metabolism of arsenic takes place in two steps. First, it starts with the reduction of As(V) to As(III) and then continues with the formation of mono-, di-, and tri-methylated products by oxidative biological methylation of As(III). However, there is a large individual variation in arsenic methylation capacity in humans. Additionally, the toxic effect of each As species depends on the associated metabolism in human bodies. When MMA or DMA are orally administered, they are excreted in the urine largely without changing chemical structure. Only a minor proportion of MMA is converted to DMA and no further metabolism of DMA is observed [1, 3, 6, 7]. Arsenosugars appear to be metabolized to several arsenic compounds including DMA, but arsenobetaine is excreted rapidly and unchanged in urine [8].

There are many studies dealing with the positive association between arsenic exposure and cancers of different organs. The inorganic arsenic in drinking water has been associated with skin cancer in humans. Moreover, dimethlyarsinic acid (DMA), a metabolite of arsenite, can induce bladder cancers at high concentrations in drinking water, but not skin cancers [9].

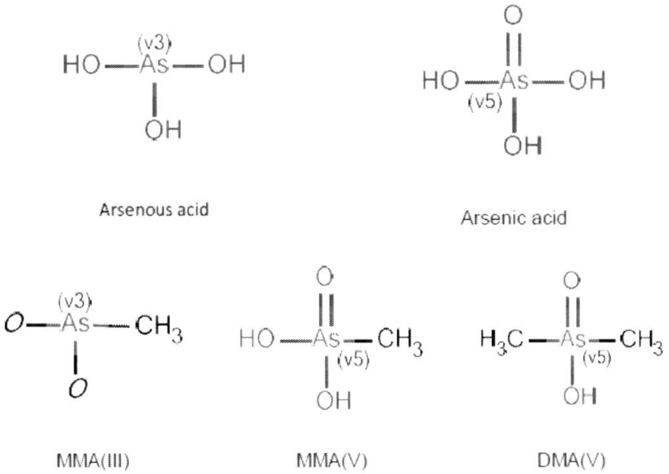

Figure 2. Molecular structures of arsenous acid, arsenic acid, monomethylarsonous acid (MMAIII), monomethylarsonic acid (MMAV) and dimethylarsinic acid (DMAV).

Furthermore, there is strong evidence that ingested arsenic causes kidney and ureter transitional cell carcinomas [10]. Also, inorganic arsenic has been associated with lung and liver cancer [2, 11, 12]. There is evidence that inter-individual differences in arsenic metabolism may be an important risk factor for arsenic-related lung cancer, and may play a role in cancer risks among people exposed to relatively low arsenic water concentrations [13]. In many cases, it has been emphasized that the cigarette smoking has a synergistic effect with the ingestion of As from drinking water to the development of cancer.

Likewise, non-carcinogenic effects can be caused by the ingestion of As, including skin lesions (e.g., hyperpigmentation, melanosis, keratosis), respiratory system problems (e.g., chronic cough, shortness of breath, bronchitis), nervous system effects (e.g., neuropathy, neurobehavioral, weakened memory, decreased attention), and reproductive effects (e.g., pregnancy complications, fetus abnormalities, premature deliveries, reduced birth weight). There are, in addition, potential links to heart disease and diabetes, but further evidence is needed to support these relationships [14].

In 2010 the Joint FAO/WHO Expert Committee on Food Additives (JECFA) organized a meeting in Italy, with the purpose to evaluate certain contaminants in food. It was concluded that the provisional tolerable weekly intake (PTWI) of 15 μg/kg body weight (equivalent to 2.1 μg/kg body weight per day) for inorganic arsenic, which was set at the thirty third meeting of the Committee, is no longer appropriate. The inorganic arsenic lower limit on the benchmark dose for a 0.5 % increased incidence of lung cancer (BMDL0.5) was determined from epidemiological studies to be 3.0 μg/kg body weight per day (2–7 μg/kg body weight per day based on the range of estimated total dietary exposure) using a range of assumptions to estimate total dietary exposure to inorganic arsenic from drinking water and food. The uncertainties in this BMDL relate to the assumptions regarding total exposure and to extrapolation of the BMDL0.5 to other populations due to the influence of nutritional status, such as low protein intake, and other lifestyle factors on the effects observed in the studied population. The Committee noted that PTWI of 15 μg/kg body weight is in the region of the BMDL0.5 and therefore it was withdrawn [15].

On the other hand, the European Food Safety Authority (EFSA) noticed that most of the food data, collected in the framework of official food control, are reported as total arsenic without differentiating the various arsenic species. This fact can lead to an overestimation of the health risk related to dietary arsenic exposure, since the inorganic As is more toxic that the organic As.

Therefore, the EFSA Panel on Contaminants in the Food Chain (CONTAM Panel) collected data from 15 European countries to assess the human exposure.

The proportion of inorganic arsenic was assumed to vary from 50 to 100 % of the total arsenic reported in food commodities other than fish and seafood, with 70 % considered as best reflecting an overall average. In fish and seafood the relative proportion of inorganic As is small and tends to decrease as the total As content increases. The CONTAM Panel concluded that the JECFA PTWI of 15 µg/kg body weight is no longer appropriate and therefore, it concluded that the overall range of the lower confidence limit of the benchmark dose of 1 % extra risk (BMDL01) values of 0.3 to 8 µg/kg body weight per day should be used instead of a single reference point in the risk characterization for inorganic arsenic [16]. In general, there is a great concern about the definition of the maximum permissible levels of total and inorganic arsenic. The European Union (EU) has adopted in the Council Directive 98/83/EC, the concentration of 10 µg/L for the arsenic in drinking water [17].

Similarly, the United States Environmental Protection Agency (US EPA) has established the zero value as the Maximum Contaminant Level Goal (MCLG) for As in drinking water, that is the level of a contaminant in drinking water below which there is no known or expected risk to health. Based on the MCLG, EPA has set an enforceable regulation for arsenic, called a Maximum Contaminant Level (MCL), at 10 µg/L, which is the highest level of a contaminant that is allowed in drinking water. MCLs are set as close to MCLGs as feasible using the best available treatment technology and taking cost into consideration. MCLs are enforceable standards [18]. The concentration of 10 µg As/L in drinking water that has been adopted by both the WHO and the US EPA is still higher than the proposed Canadian and Australian maximum permissible concentrations of 5 and 7 µg As/L, respectively [14]. Finally, the European Commission in the Commission Regulation (EU) No 744/2012 has established limits for As in products intended for animal feed [19].

Despite the fact that the toxicity of total As and/or inorganic As is well known, there are no European Union, United States or WHO maximum limits in foodstuffs [20, 21]. On the contrary, the Food Standards Australia New Zealand (FSANZ) agency has established the maximum levels (MLs) of total arsenic and inorganic arsenic in some food categories. However, it is noted in the Standard 1.4.1 – Contaminants and Natural Toxicants (F2013C00140) that "regardless of whether or not a maximum level exists, the levels of

            I. N. Pasias, A. K. Psoma, N. I. Rousis et al.

contaminants and natural toxicants in all foods should be kept As Low As Reasonably Achievable (the ALARA principle)". The maximum level of total arsenic in cereals is 1 mg/kg and the respective of inorganic arsenic in crustaceans and fishes is 2 mg/kg and in mollusks and seaweed is 1 mg/kg [22, 23].

Table 1. Indicators of Maximum Levels of Arsenic in Foods adopted by China

Food type / Name	Maximum Level (MLs) / (mg/kg)	
	Total arsenic	Inorganic arsenic
Grains and their products (excluding paddy and rice)	0.2	-
Paddy and rice	-	0.2
Aquatic products		
Fish	-	0.1 (by fresh weight)
Shellfishes, crustaceans, cephalopods and other aquatic products	-	0.5 (by fresh weight)
Vegetables	0.5	-
Edible fungi		
Edible fungus	0.5	-
Dried edible fungus	1	-
Meat and meat products	0.5	-
Milk and milk products		
Fluid milk (raw milk, pasteurized milk, sterilized milk, acidified milk and modified milk)	0.1	-
Milk powder	0.5	-
Fats and oils, and fat emulsions	0.1	-
Condiments	0.5	-
Sweeteners		
Sugar	0.5	-
Beverages		
Packaged drinking water	0.01	-
Cocoa products, chocolate and chocolate products and candies	0.5	-
Special nutritious foods		
Cereal supplementary foods for infant and babies (excluding products with algae)	-	0.2
Cereal supplementary foods with algae for infant and babies	-	0.3
Canned supplementary foods for infant and babies (excluding the products made from aquatic products and animal liver)	-	0.1
Canned supplementary foods made from aquatic products and animal liver for infant and babies	-	0.3

Finally, on August 2012, China's Ministry of Health notified the World Trade Organization (WTO) of Maximum Levels of Contaminants in Foods as G/SPS/N/CHN/312. China has one of the most stringent regulations for total As and inorganic As, especially in rice. In Table 1 are summarized all the maximum levels of total As and inorganic As for different food types, that have been adopted by China [24].

3. CONTENT OF INORGANIC ARSENIC IN DIFFERENT FOODSTUFFS

As already discussed in the previous section, arsenicis presented as an important aquifer contaminant of geogenic origin worldwide (Figure 1). Extremely elevated concentrations of As are detected in large deltas and along major rivers in developing countries of South and East Asia (the Bengal Delta Plain in West Bengal, India, Bangladesh, Mekong Delta in Cambodia and the Red River Delta in Vietnam) [3]. In addition high As concentrations have been reported from several parts of the United States such as, California, Alaska, Arizona, Indiana, Idaho, Nevada, Washington, Missouri, Ohio, Wisconsin, and New Hampshire. Natural occurrences of As have also been found in Canada, Argentina, Mexico, Chile, Taiwan, China, Japan, Thailand, Ghana, Hungary, and United Kingdom (Figure 3) [14].

One of the most principal human exposures to inorganic As is through groundwater. Conservative estimations conclude that more than 100 million people, especially in developing countries, are at risk of inAs exposure [3]. Halder et al. (2013) in a recent research in rural Bengal concluded that local people are exposed to high concentrations of inorganic As largely vary from below the instrumental limit of detection (LOD) (<1 µg/L) to 875 µg/L. Only the 37.5% of the water sources in this study is appeared to be safe (<10 µgAs/L). In the 25% and 37.5% of the tube wells the As concentration was found to be >10−50 and >50µg/L, respectively [25]. Furthermore, a study from four districts in the Red River Delta, Vietnam, conducted by Agusa et al. (2013), showed As concentrations in groundwater ranging from <1 to 632 µg/L. Even though sand filters are used for the removal of As in groundwater with great average As removal efficiencies between 85-91% there were still water samples that exceeded the permissible value set by WHO and also Vietnamese Ministry of Health [26]. In Hungary, Sugáret al. (2013) reported a total As concentration above the established limit value of 10 µg/L in twenty two out of twenty three water samples from public wells collected from three different areas, ranging between 7.2-210.3 µg/L [27]. However, in the first total diet study from Hong Kong (China), it was found that no water sample was found to contain detectable inorganic As [28].

The contamination of food with elevated As concentrations is another aspect of human exposure to inorganic As. Food products like vegetables, rice and cereals which is staple food, especially for the population in developing countries are inorganic As contaminated since the cultivations are irrigated with As-contaminated groundwater [14]. The dietary exposure to inorganic As increased apart from the contaminated drinking water by the daily consumption of rice, rice flour, and rice-based products (e.g., rice noodles), cereal and cereal-based products, seafood, fishes, fish oil and vegetables. The risk for the children under three years old is 2-3 folds that of adults since their dietary exposure to inorganic As is directly related to the intake of rice flour and cereal-based infant products [29]. As already discussed, international maximum levels for inorganic As in foodstuffs have not been established yet neither in EU nor within the framework of Codex Alimentarius contrary to some other countries, namely China (Table 1), Australia and New Zealand (1–2 mg/kg for seafood products) [3, 22, 23]. The Codex Alimentarius Committee proposed a draft maximum level of 300 µg/kg and 200 µg/kg for raw and polished rice, respectively [30].

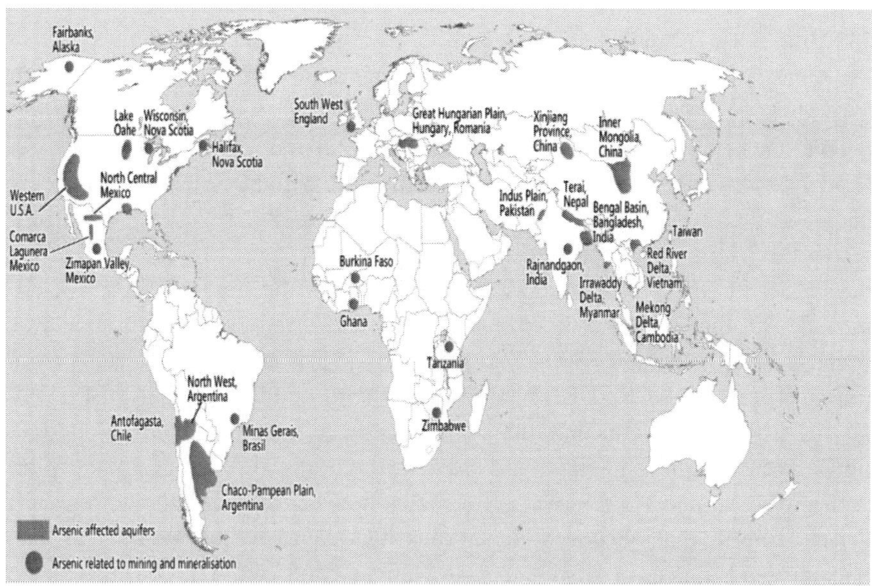

Figure 3. Arsenic occurrences in groundwater from selected parts of the world.

Rice (*Oryza sativa*) is considered as major staple food globally and as a consequence is one of the most significant sources of dietary intake of inorganic As. The daily intake in Asian countries reaches up to 0.5 kg (dry weight) per person and the inorganic As varying from 10 to 90% of the total arsenic [31]. The high mobility of As under the flooding conditions that rice is cultivated makes it capable to As accumulation [21]. In a recent study by Sofuoglu et al. (2014), the average inorganic As concentrations in rice and bulgur samples collected from 50 participants living in Izmir, Turkey ranging from 0.013 to 0.393 (µg/kg) for rice and <0.001 to 0.72 (µg/kg) for bulgur, corresponding to 1-15% and 0-3.6% of the inorganic As Tolerable Daily Intake (TDI) for rice and bulgur, respectively [32]. Rasmussen et al. (2013) calculated the inorganic As concentrations in various types of rice products samples (n=36).

The white rice contained a median value of 0.09 mg/kg, whereas for black and brown rice the concentrations were very low, both at 0.14 mg/kg. The rice crackers were those that contained the highest inorganic As concentration (median 0.28 mg/kg and up to 0.57 mg/kg) along with some individual samples of black and brown rice that contained up to 0.3–0.4 mg/kg exceeding the MLs for raw rice (0.3 mg/kg) proposed by Codex [33]. In a recent published study conducted by Pasias et al. (2013) among eighteen different types and brands of rice and rice flour purchased from Greek market, authors stated that the proportion of total inorganic arsenic was equal to (64±19)% of the total, ranging between 30.1 to 147 µg/kg. The lowest inorganic As concentration was calculated for Basmati rice from India on the contrary to parboiled rice from Greece which showed the highest concentration [34]. A similar study was carried out with eight different brands of long grain rice and ten different brands of baby porridge powders that were composed only on rice or rice and other cereals purchased from Finnish supermarket by Rintala et al. (2014). The findings showed that the inorganic As concentration among long grain rice ranged from 0.09 to 0.28 mg/kg, with an average at 0.16 mg/kg which represented the 74% of total arsenic and for the baby porridge powders was between 0.07 to 0.21 mg/kg [35]. Table 2 shows a summary of total As and inorganic As concentrations in different rice varieties and rice-products from selected countries.

In many countries that the diet is not rise-based, wheat is the most consumed grain. On this basis, Llorente-Mirandes et al. (2014) published a very detailed research studying 30 cereal-based foods, representing all the types of cereal products that are widely consumed in Barcelona, Spain. The highest inorganic As concentrations apart from rice (74.3 µg/kg) was found in three different samples of infant cereals with an average of 23.5µg/kg. Among those with the lowest concentrations were two samples of corn-based snacks (3.1-3.6 µg/kg), chocolate cookies (3.8 µg/kg), corn-based breakfast cereals (3.3 µg/kg) and wheat flour (3.9 µg/kg).

Table 2. Total As and inorganic As concentrations in different rice varieties and rice-products from selected countries of origin

Origin	Rice variety/product	n [a]	Total As (mg/kg)	inAs (mg/kg)	% inAs	Ref.
Italy	Arborio	29	0.23	0.10	25	[36]
	Carnaroli	17	0.23	0.09	30	
	Ribe	21	0.18	0.08	22	
	Ribe/Roma parboiled	10	0.20	0.11	11	
	Roma	8	0.19	0.08	36	
	VialoneNano	5	0.28	0.11	21	
	Originario	3	0.19	0.10	21	
	Others (mixture varieties)	8	0.18	0.10	19	
USA	Medium-grain rice	3	-	0.074	-	[37]
	Enriched long-grain rice	3	-	0.074	-	
	Brown medium-grain rice	3	-	0.094	-	
	OrganicHaiga medium-grain rice	3	-	0.083	-	
	Wild black rice	3	-	0.045	-	
	Glutinous rice	3	-	0.066	-	
Thailand	Jasmine rice	3	-	0.077	-	
	Wild red rice	3	-	0.086	-	
	Matta rice	3	-	0.049	-	
	Basmati rice	3	-	0.060	-	
India	Brown Basmati rice	3	-	0.235	-	
	Ponni rice	3	-	0.057	-	
	Organic Basmati rice	3	-	0.046	-	
Vietnam	Jasmine rice	3	-	0.066	-	
China	infant rice	14	0.135	0.114	76	[29]
USA	infant rice	5	0.253	0.125	55	
UK	infant rice	5	0.237	0.162	71	
Spain	infant rice	7	0.181	0.085	53	
Thailand	Thai Jasmine rice	1	0.127	0.120	-	[38]
	Bio quality Thai bonnet natural	1	0.211	0.182		
Italy	Long grain natural Lagris	1	0.108	0.077	-	
	Bio long grain natural Harmony	1	0.218	0.171	-	
	Long grain Menu	1	0.087	0.058	-	
Pakistan	Basmati F.W. Tandoori	1	0.049	0.030	-	
India	Tesco Basmati	1	0.093	0.072	-	
EU	Basmati Uncle Bens	1	0.036	0.025	-	
	Long grain Uncle Bens	1	0.129	0.077	-	
Poland	Long grain white rice Euro shopper	1	0.157	0.114	-	

[a] number of samples that were analysed.

The average concentrations of inAs was found to be 5.9 µg/kg for bread, 5.2 µg/kg for biscuits 6.4 µg/kg for breakfast cereals, 3.1 µg/kg for flour, 4.5 µg/kg for snacks, 12.7 µg/kg for pasta and 16.6 µg/kg for infant cereals [39].

The concentrations of inorganic As in seafood are significantly lower than the grain products and vary depending on the type, because arsenic is mainly found as non-toxic organic arsenic species. The EFSA have set fixed values for inorganic As of 0.03 mg/kg in fish and 0.1 mg/kg in seafood [16]. In the past few years much research has been conducted, studying the inorganic As concentration in different types of fishes, seafood, canned fishes and fish oil products.

The interest focuses on areas with elevated As concentrations such as Taiwan. There, Lianget al. (2013) estimated the median inorganic As intakes for four different seafood that were 0.102, 0.520, 0.003, 0.087, and 0.745 µg/day for tilapia, milkfish, mullet, oyster, and clams, respectively. Moreover, in a Mediterranean study by Copat et al. (2013) the daily intake of inorganic As per meal size was estimated for different types of fishes for an adult and for a child and compared with tolerable intake (µg/kg-daily), as suggested by FAO/WHO JECFA. The estimations were made by assuming that the inorganic As was equal to 3% of the total As concentration. The highest daily intake of inorganic As is achieved by the consumption of goatfishes (*M. barbatus*) both for an adult and for a child with 1.073 and 2.357 µg/kg-daily, respectively.

The lowest daily intake of inorganic As was estimated for the consumption of saltwater clams (*D. trunculus*) ranging from 0.148 to 0.326 µg inorganic As/kg-daily [41]. In the interesting study of Ruttens et al. (2012), who determined the inorganic As concentrations in ninety eight samples of fish, mollusks and crustaceans purchased from Belgian market, only scampi, shrimps, and mussels were these products in which inorganic As could be quantified (Limit of Quantification or LOQ: 0.001 mg/kg for As(III) and 0.002 mg/kg for As(V)) with concentrations ranged from 0.005 to 0.022 mg/kg [42]. With a great concern in the inorganic As intake of specific population groups such as children, López-García et al. (2011) calculated the levels of inorganic As in fish-based baby foods. The results showed that no inorganic As was found for all the seven different fish-based baby foods analyzed [43].

Finally, since a lot of interest nowadays has been shown in the consumption of the so called "dietary supplements", two very recent studies calculated the proportion of inorganic As in dietary supplements purchased from Mexican [44] and Danish market [45].

Table 3. A summary of the daily inorganic arsenic intake from different food commodities in China, Japan and European Union (EU)

	China[46]			Japan[47]			E.U[48]		
	inAs (mg/kg)	Food Intake (g/day)	Daily inAs intake (µg/d)	inAs (mg/kg)	Food Intake (g/day)	Daily inAs intake (µg/d)	inAs (mg/kg)	Food Intake (g/day)	Daily inAs intake (µg/d)
Rice									
White rice	0.103	238	24.5	-	-	-	0.072-0.162	175	0.22
Brown rice								175	0.38
Flour, Bread	0.015	140	2.10	-	-	-	0.000-0.026	290	0.06
Cereals	0.060	23.6	1.42	0.032	424	13	0.0004-0.046		
Pulses	0.036	16.0	0.58	nd	50.4	0.008	0.001-0.032		
Vegetables	0.025	276	7.02	nd	276	0.014	0.0003-0.057		
Fruits	0.014	45.0	0.61	nd	112	0.017	0.000-0.031		
Meat	0.024	78.6	1.89	nd	80.3	0.028	0.001-0.020		
Milk	0.021	26.6	0.56	nd	130	0.019	0.001-0.025	900	0.05
Dairy products							0.000-0.027		
Eggs	0.018	23.7	0.43	nd	36.2	0.007	0.003-0.009		
Seafood	0.112	29.6	5.03	0.003	74.6	0.200	0.001-0.131	180	0.03
Potatoes				0.009	62.2	0.540	0.002-0.007		
Sugar & sweeteners				<0.002	7.20	0.007	0.000-0.032		
Nuts & seeds				<0.002	3.00	0.003	0.00-0.047		
Mushrooms				0.005	16.6	0.078	0.007-0.221		
Algae				0.653	12.7	8.30	6.13-6.14		
Oils & fats				<0.002	10.0	0.010	0.002-0.045		
Pastries				0.011	27.5	0.310	0.0002-0.067		
Beverages				nd	627	0.270	0.003-0.011	1350	0.13
Seasonings & spices				nd	98.5	0.020	0.000-0.268		
Supplemental nutrient				0.013	16.5	0.210	0.0003-1.49		

	China[46]			Japan[47]			E.U[48]		
	inAs (mg/kg)	Food Intake (g/day)	Daily inAs intake (μg/d)	inAs (mg/kg)	Food Intake (g/day)	Daily inAs intake (μg/d)	inAs (mg/kg)	Food Intake (g/day)	Daily inAs intake (μg/d)
food									
Drinking water									
Tap water				nd	600	0.290	0.001-0.002	2500	0.08
Beer							0.003-0.011	2600	0.25

The supplements from Mexico showed that the concentrations of inorganic As ranged from 0.14 to 0.28 mg/kg dry weight and the daily intake was calculated to range from 0.21 to 0.83 µg/day [44], whereas in the Danish research the inorganic As concentrations ranged from 0.03 mg/kg, for *Chlorella pyrenoidosa* algae capsules, to 3.2 mg/kg for a herbal supplement of Chinese origin containing andrographis, dandelion and woad [45]. Table 3 shows a summary of the daily inorganic arsenic intake from different food commodities in China, Japan and European Union (EU).

4. ANALYTICAL METHODS FOR THE DETERMINATION OF INORGANIC ARSENIC

The levels of inorganic arsenic are not routinely monitored since there is no directive regarding its determination in different foodstuffs. In general, the analytical procedures followed for the determination of inorganic arsenic must be precise and accurate with appropriate limits of detection corresponding to the content in the foodstuffs and the crucial values set from the different organizations [2, 16, 18, 19, 22, 24]. The main steps followed for the determination of inorganic arsenic are (i) the preparation of the samples, (ii) the appropriate extraction procedure for the determination of specific forms of the arsenic, and (iii) its determination of using the appropriate technique.

In general there are two main approaches for the arsenic speciation analysis: (a) the separation of arsenic species after chromatographic separation and detection with specific detectors. These methods usually use different chromatographic techniques, such as gas (GC) or liquid (LC) chromatography with on-line detection techniques, such as mass spectrometry (MS), inductively coupled plasma mass spectrometry (ICP-MS), atomic absorption spectrometry (AAS), or atomic fluorescence spectrometry (AFS) and (b) the separation of arsenic species after the selective extraction(s) based on the different chemical properties of each of the species. These methods use mainly spectrometric techniques and chemical methods for the separation of one species from the others [1, 20, 23, 31, 34, 49-53]. Electrothermal atomic absorption spectrometry (ETAAS), hydride generation atomic absorption spectrometry (HG-AAS), inductively coupled plasma – mass spectrometry (ICP-MS) and high performance liquid chromatography coupled with inductively coupled plasma – mass spectrometry (HPLC-ICP-MS) are among the most used techniques. However, only few methods are fully validated, concerning all those criteria such as precision, accuracy, and participation in interlaboratory schemes [34], and even these works are not capable for application in all matrices. Therefore, recently, the Directorate General for Health and Consumers (DG SANCO) of the European Commission (EC) recently requested that the European Union Reference Laboratory for Heavy Metals in Feed and Food (EU-RL-HM) to evaluate the performance of European laboratories focused on the determination of total and inorganic As in rice with a view to future discussions on the need for possible regulatory measures [20].

Hydride generation atomic absorption spectrometry (HG-AAS) is widely used for the determination of inorganic arsenic species, since it provides high sensitivity and selectivity and low limits of detection. Differentiation of arsenic(V) and arsenic(III) is achieved by adjusting the pH of the acid reacting with $NaBH_4$. The main preparation procedure steps are:(a) the extraction of the arsenic species using nitric or hydrochloric acid and different extraction systems such as, sonication or microwave-assisted extraction, (b) the reduction of As(V) using KI, ascorbic acid and hydrochloric acid and (c) the use of the hydride generation system in order to determine selectively the As(III) and As(V) contents [20, 54-57].

Flow injection coupled with hydride generation atomic absorption [58, 59] based on dispersive liquid-liquid microextraction (DLLME) has also been developed. In this method, As(III) was complexed with ammonium pyrrolidinedithiocarbamate at pH 4 Then, As(III) was extracted into the ionic liquid. Recently, Rasmussen et al. (2012, 2013) have developed a solid-phase extraction procedure using strong anion exchange SPE cartridge enabled the selective inorganic As quantification by HG-AAS, measuring total arsenic (As) in the SPE elute after solubilized and oxidized all inorganic As to As(V) using HNO_3 and H_2O_2[33, 60].

Electrothermal Atomic absorption Spectrometry (ETAAS) is also presented as a potential technique for not only the determination of the total inorganic arsenic but also for the selective determination of As(III) and As(V) species. This technique provides similar limits of detection with HG-AAS, but is mainly used for the speciation of inorganic arsenic in water samples after the complexation of As(III) with ammonium pyrrilidinedithiocarbomate at pH=4 [61]. Recently, ETAAS has also been used for the separation and the accurate determination of As(III) and As(V) in other matrices such as muscle tissues of fish species and rice or rice flour samples [34, 62].

A coupled system including HPLC and ICP-MS or ETAAS gives a suitable method for the determination of non-volatile arsenic species, providing low limits of detection and high sensitivity and selectivity [61]. The major drawbacks for this technique are: (a) the incompatibility of the common mobile phases with the atomization systems and plasma sources [55, 56, 61], (b) the use of low organic content mobile phase does not allow the determination of non-polar arsenic species [57]. Ion exchange or reverse phase columns are mainly used for the chromatographic separations, but it was proved that the ion exchange based systems enhance the separation and the selectivity, since the analytes interact directly with the stationary phase are therefore less prone to interferences from co-chromatographed matrix constituents [61, 63-68]. The extraction of inorganic arsenic is also based on the use of microwave-assisted or sonication systems after the addition of diluted nitric or hydrochloric acid and hydrogen peroxide, as described by the laboratories that participated in the proficiency tests organized by the EU-RL-HM on the determination of inorganic arsenic in rice, wheat, vegetable food and algae [20, 54]. Gas chromatography, either coupled with mass spectrometry either coupled with ICP-MS, provides excellent separation and detection, but is only applicable for volatile arsenic species, and for this reason it is rarely used for the determination of inorganic arsenic [57].

Voltammetric and polarographic techniques have also been used for the speciation of arsenic with a detection limit of 0.2 ppb, which is totally acceptable for the majority of food matrices. Mays and Hussam (2009) had presented an excellent review concerning an overview of all voltammetric techniques used for the speciation of arsenic since 2001 [69].

All methods presented in the literature for the determination of inorganic arsenic in food commodities are of great concern, since there are matrix-dependent. Even between expert laboratories there is a great variance among the values given for the different food matrices. For example, in the 12th proficiency test organized by the EU-RL-HM (IMEP-112) that focused on the determination of total and inorganic arsenic in wheat, vegetable food and algae, two certified expert laboratories were excluded from the final report, since some systematic errors were presented in their results [54]. Furthermore, the results as reported by the participants, regarding the z and ζ scores, for inorganic As, about 60 and 75 % of the participants reported satisfactory results for wheat, rice flour and vegetable food, respectively, despite the relatively low concentration of inorganic As in vegetable food. However, <20 % of the participants scored satisfactorily for inorganic As in algae. The distribution of satisfactory results reported for the test materials included in this exercise could reflect the difficulty introduced by the different matrices. Finally, these proficiency tests also indicated the lack of experience on the determination of inorganic arsenic since the number of laboratories that determined inorganic As was considerably lower than the number of laboratories that determined total As, and that the majority of the laboratories reported less than 50 inorganic As determinations per year [20, 54].

Taking into account all studies reported in the literature, certain quality assurance criteria must be followed during the procedure. The method followed for inorganic arsenic determination in food commodities must be an in-house developed method or a standard reference method. There are only few official methods concerning the determination of inorganic arsenic in several matrices, such as, BVL L 25.06-1:2008-12 for the determination of inorganic arsenic in algae by HGAAS, EN 15517:2008 for the determination of inorganic arsenic in seaweed by HGAAS, and EN 16278:2012 for the determination of inorganic arsenic animal feeding stuffs by HGAAS after microwave extraction and separation by solid phase extraction (SPE).

For the applicability of the developed "in house" methods, the accepted criteria for analytical method validation should be followed. Accuracy, precision, calibration curves, linearity, methods detection limits, selectivity, robustness, and uncertainty calculation must be defined in order to check the method's performance. Several certified reference materials, such as DORM-2 (Dogfish muscle), DOLT-3 (Dogfish liver), DORM-3 (Fish protein), DOLT-4 (Dogfish liver), and TORT-2 (Lobster hepatopancreas) from the National Research Council of Canada (CNRC), BCR 627 (Tuna fish tissue) from the Institute for Reference Materials and Measurements (IRMM), and 1568a (Rice Flour) from the National Institute of Standards and Technology (NIST) are available and can be used for the validation of the analytical methods. Interlaboratory proficiency tests under the auspices of organizations can also provide documented intra- and interlaboratory method performance, but are only undertaken after a thorough in-house validation of the method.

CONCLUSION

The risk assessment of the inorganic arsenic content in different foodstuffs and especially for the more sensitive population groups, such as children, and the most known developed analytical methods for its determination was described in the current chapter. It was proved that drinking water, rice, rice flour, cereals, dietary supplements, and seafood are the main sources of inorganic arsenic. This chapter has also designated the necessity of establishing a directive concerning its determination in different foodstuffs and that the modern analytical chemistry can provide useful information not only for the content of the inorganic arsenic in the foodstuffs, but also in the deeper understanding of the degradation processes. Consequently, in the future it is expected to explore new methods of arsenic speciation research dealing with the determination of all inorganic and organic arsenic species. Research in these areas should be intensified to overcome the current limitations.

REFERENCES

[1] J.–H. Huang, P. Fecher, G. Ilgen, K.–N. Hu and J. Yang, *Food Chem.* 130, 453 (2012).

[2] International Agency for Research on Cancer, *Monographs on the Evaluation of Carcinogenic Risks to Humans, Arsenic and Arsenic Compounds,* 100C (2012) http://monographs.iarc.fr/ENG/Monographs/vol100C/mono100C-6.pdf.

[3] K. Sharma, J. C. Tjell, J. J. Sloth and P. E. Holm, *Appl. Geochem.* 41, 11 (2014).

[4] J. P. Wang, L. Qi, M. R. Moore and J. C. Ng, *Toxicol. Lett.* 133, 17 (2002).

[5] K. T. Kitchin and R. Conolly, *Chem. Res. Toxicol.* 23 (2), 327 (2010).

[6] M. A. García-Sevillano, M. Contreras-Acuña, T. García-Barrera, F. Navarro and J. L. Gómez-Ariza, *Anal. Bioanal. Chem.* 406, 1455, (2014).

[7] M. A. Rahman, H. Hasegawa and R. P. Lim, *Environ. Res.* 116, 118 (2012).

[8] K. A. Francesconi, R Tanggaard, C. J. McKenzie and W. Goessler, *Clin. Chem.* 48:1, 92 (2002).

[9] T. G. Rossman, A. N. Uddin and F. J. Burns, *Toxicol. Appl. Pharmacol.* 198, 394 (2004).

[10] C. Ferreccio, A. H. Smith, V. Duran, T. Barlaro, H. Benítez, R. Valdés, J. J. Aguirre, L. E. Moore, J. Acevedo, M. I. Vasquez, L. Perez, Y. Yuan, J. Liaw, K. P. Cantor and C. Steinmaus, *Am. J. Epidemiol.* 178 (5), 813 (2013).

[11] Celik, L. Gallicchio, K. Boyd, T. K. Lam, G. Matanoski, X. Tao, M. Shiels, E. Hammond, L. Chen, K. A. Robinson, L. E. Caulfield, J. G. Herman, E. Guallar and A. J. Alberg, *Environ. Res.* 108, 48 (2008).

[12] N. Sawada, M. Iwasaki, M. Inoue, R. Takachi, S. Sasazuki,T. Yamaji, T. Shimazu and S. Tsugane, *Cancer Cause Control* 24, 1403 (2013).

[13] D. Melak, C. Ferreccio, D. Kalman, R. Parra, J. Acevedo, L. Perez, S. Cortes, A. H. Smith, Y. Yuan, J. Liaw and C. Steinmaus, *Toxicol. Appl. Pharmacol.* 274, 225 (2014).

[14] S. Kapaj, H. Peterson, K. Liber and P. Bhattacharya, *J. Environ. Sci. Health, Part A* 41, 2399 (2006).

[15] Food and Agriculture Organization of the United Nations (FAO), Rome, Italy JECFA/72/SC (2010).

[16] Scientific Opinion on Arsenic in Food EFSA Panel on Contaminants in the Food Chain (CONTAM) European Food Safety Authority (EFSA), Parma, *Italy Summary: EFSA Journal* 2009; 7(10):1351.

[17] European Union, *Council Directive* 98/83/EC.

[18] United States Environmental Protection Agency (US EPA), *Water: Basic Information about Arsenic in Drinking Water*http://water.epa.gov/drink/contaminants/basicinformation/arsenic.cfm.

[19] European Commission, *Commission Regulation* (EU) No 744/2012.

[20] M. B. de la Calle, H. Emteborg, T. P. J. Linsinger, R. Montoro, J. J. Sloth, R. Rubio, M. J. Baxter, J. Feldmann, P. Vermaercke and G. Raber, *Trends Anal. Chem.* 30 (4), 641 (2011).

[21] Y.–G. Zhu, P. N. Williams, A. A. Meharg, *Environ. Pollut.* 154, 169 (2008).

[22] Food Standards Australia New Zealand (FSANZ), *Australia New Zealand Food Standards Code - Standard 1.4.1 - Contaminants and Natural Toxicants - F2013C00140*http://www.comlaw.gov.au/Details/F2013C00140.

[23] P. N. Williams, A. H. Price, A. Raab, S. A. Hossain, J. Feldmann and A. A. Meharg, *Environ. Sci. Technol.* 39, 5531 (2005).

[24] Ministry of Health, Beijing, China, *National Food Safety Standard - Maximum Levels of Contaminants in Food* - G/SPS/N/CHN/312.

[25] D. Halder, S. Bhowmick, A. Biswas, D. Chatterjee, J. Nriagu, D.N.G. Mazumder, Z. Šlejkovec, G. Jacks and P. Bhattacharya, *Environ. Sci. Technol.* 47, 1120 (2013).

[26] T. Agusa, P.T.K.Trang, V.M. Lan, D.H. Anh, S. Tanabe, P.H. Viet and M. Berg, *Sci. Total Environ.* in press (2013), http://dx.doi.org/10.1016/j.scitotenv.2013.10.039.

[27] É.Sugár, E. Tatár, G. Záray and V.G. Mihucz, *Microchem. J.* 107, 131 (2013).

[28] S. W.-C. Chung, C.-H Lam and B. T.-P. Chan, *Food Addit. Contam Part A*(2014), DOI: 10.1080/19440049.2013.877162.

[29] Á.A. Carbonell-Barrachina, X. Wu, A. Ramírez-Gandolfo, G.J. Norton, F. Burló, C. Deacon and A.A. Meharg, *Environ. Pollut.*163, 77 (2012).

[30] WHO/FAO (2012) Proposed Draft Maximum Levels for Arsenic in Rice (at step 3). Accessed 08 March 2014.

[31] Y.J. Zavala, R. Gerads, H. Gürleyük and J. M. Duxbury, *Environ. Sci. Technol.* 42, 3861 (2008).

[32] S.C. Sofuoglu, H. Güzelkaya, Ö. Akgul, P. Kavcar, F. Kurucaovalı and A. Sofuoglu, *Food Chem. Toxicol.* 64, 184 (2014).

[33] R.R. Rasmussen, Y. Qian and J.J. Sloth, *Anal. Bioanal. Chem.* 405, 7851 (2013).

[34] I.N. Pasias, N.S. Thomaidis and E.A. Piperaki, *Microchem. J.* 108, 1 (2013).

[35] E.M. Rintala, P. Ekholm, P. Koivisto, K. Peltonen and E.R. Venäläinen, *Food Chem.* 150, 199 (2014).

[36] Sommella, C. Deacon, G. Norton, M. Pigna, A. Violante and A.A. Meharg, *Environ. Pollut.* 181, 38 (2013).

[37] G. Chen and T. Chen, *Talanta* 119, 202 (2014).

[38] G. Raber, N. Stock, P. Hanel, M. Murko, J. Navratilova and K.A. Francesconi, *Food Chem.* 134, 524 (2012).

[39] T. Llorente-Mirandes, J.Calderón, F. Centrich, R. Rubio and J.F. López-Sánchez, *Food Chem.* 147, 377 (2014).

[40] C.Ping Liang, C. S. Jang, J.S. Chen, S. W. Wang, J. J. Lee and C. W. Liu, *Environ. Geochem. Health* 35, 455 (2013).

[41] C. Copat, G. Arena, M. Fiore, C. Ledda, R. Fallico, S. Sciacca and M. Ferrante, *Food Chem. Toxicol.* 53, 33 (2013).

[42] Ruttens, A. C. Blanpain, L. De Temmerman and N. Waegeneers, *J. Geochem. Explor.* 121, 55 (2012).

[43] López-García, M.Briceño and M. Hernández-Córdoba, *Anal. Chim. Acta* 699, 11 (2011).

[44] L. García-Rico and L. Tejeda-Valenzuela, *Environ.Monit. Assess.* 185, 6111 (2013).

[45] R.V. Hedegaard, I.Rokkjær and J.J. Sloth, *Anal. Bioanal. Chem.* 405, 4429 (2013).

[46] G. Li, G. X. Sun, P. N. Williams, L. Nunes and Y. G. Zhu, *Environ. Int.* 37, 1219 (2011).

[47] T. Oguri, J. Yoshinaga, H. Tao and T. Nakazato, *Arch. Environ. Contam. Toxicol.* 66, 100 (2014).

[48] *EFSA Journal* 12:3 (2014) http://www.efsa.europa.eu/en/efsajournal/doc/3597.pdf.

[49] S. Torres-Escribano, M. Leal, D. Vélez and R. Montoro, *Environ. Sci. Technol.* 42, 3867 (2008).

[50] D.J. Butcher, *Appl. Spectrosc. Rev.* 42, 1 (2007).

[51] S. Stasinakis and N.S. Thomaidis, *Crit. Rev. Environ. Sci. Technol.* 40, 307 (2010).

[52] S.-H. Nam, H.-J. Oh, H.-S. Min and J.-H. Lee *Microchem. J.* 95, 20 (2010).

[53] J. Koh, Y. Kwon and Y.-N. Pak, *Microchem. J.* 80, 195 (2005).

[54] M. B. de la Calle, I. Baer, P. Robouch, F. Cordeiro, H. Emteborg, M. J. Baxter, N. Brereton, G. Raber, D. Velez, V. Devesa, R. Rubio, T. Llorente-Mirandes, A. Raab, J. Feldmann, J. J. Sloth, R. R. Rasmussen, M. D'Amato and F. Cubadda, *Anal. Bioanal. Chem.* 404, 2475 (2012).

[55] M. Moritáand J. S. Edmonds, *Pure & Appl. Chem.*64, 575 (1992).

[56] C. K. Jain and I. Ali, *Wat. Res. 34*, 4304 (2000).

[57] K. A. Francesconiand D. Kuehnelt, *Analyst*, 129, 373 (2004).

[58] H. Shirkhanloo, A. Rouhollahi and H. Z. Mousavi, *Bull. Korean Chem. Soc., 32* 3923 (2011).

[59] O. Díaz, Y. Tapia, O. Muñoz, R. Montoro, D. Velez and C. Almela. *Food Chem. Toxicol. 50,*744 (2012).

[60] R. R. Rasmussen, R. V. Hedegaard, E. H. Larsen and J. J. Sloth, *Anal. Bioanal. Chem.* 403, 2825(2012).

[61] D. Q. Hung, O. Nekrassova and R. G. Compton, *Talanta*, 64 269 (2004).

[62] Q. Shah, T. G. Kazi, J. A. Baig, M. B. Arain, H. I. Afridi, G. A. Kandhro, S. K. Wadhwa and N. F. Kolachi, *Food Chem.*119, 840 (2010).

[63] D. Beauchemin, K.W.M. Siu, J. W. McLaren and S.S. Berman, *J. Anal. At. Spectrom.* 13,285 (1989).

[64] T. van Elteren and Z. Slejkovec, *J. Chromatogr. A* 789 339 (1997).

[65] K. Falk and H. Emons, *J. Anal. At. Spectrom.* 15,643 (2000).

[66] B. He, G.-B. Jiang, X.-B. Xu, *Fresenius J. Anal. Chem.* 368,803 (2000).

[67] S. Miyashita, M. Shimoya, Y. Kamidate, T. Kuroiwa, O. Shikino, S. Fujiwara, K. A. Francesconi, and T. Kaise, *Chemosphere* 75, 1065 (2009).

[68] C. P. Verdon and K. L. Caldwell, *Anal. Bioanal. Chem.* 393, 939 (2009).

[69] D. E. Mays and A. Hussam. *Anal. Chim. Acta* 646, 6 (2009).

In: Arsenic
Editor: Marissa Jane Olson

ISBN: 978-1-63321-054-7
© 2014 Nova Science Publishers, Inc.

Chapter 5

ARSENIC EXPOSURE INHIBITS POLY (ADP)-RIBOSYLATION OF NUCLEAR PROTEINS IN PC-12 CELLS AND RAT BRAIN

*Fátima Ceballos[1], Juan Manuel Delgado[1], Claudia G. Castillo[1], Othir Galicia-Cruz[2] and María E. Jiménez-Capdeville[*1]*
[1]Departamento de Bioquímica, Facultad de Medicina, and
[2]Departamento de Farmacología, Facultad de Medicina, Universidad Autónoma de San Luis Potosí, San Luis Potosí, SLP, México

ABSTRACT

Context: Recent reports demonstrate that poly (ADP)-ribosylation participates in learning processes, together with other epigenetic mechanisms of chromatin remodeling. The enzyme poly (ADP-ribose) polymerase 1 (PARP-1) has zinc finger domains that are target of inhibition by arsenic, therefore inhibition of this enzyme in the central nervous system may contribute to the cognitive deficits observed in humans after prolonged low level arsenic exposure, especially during

---

[*] Corresponding Author, María E. Jiménez Capdeville, Ph.D., Departamento de Bioquímica, Facultad de Medicina, UASLP., Ave. Venustiano Carranza No. 2405, 78210 San Luis Potosí, SLP, México., Phone and FAX: +52 (444) 826 2300 ext.6630, E-mail: mejimenez@uaslp.mx.

development. *Objective:* The aim of this work was to evaluate *in vitro* and *in vivo* PARP-1 inhibition by arsenite. *Materials and methods:* PARP-1 activity was analyzed by quantification of substrate concentration (NAD+) and immunoreactivity for the reaction products (poly (ADP-ribosylated) proteins) in arsenic exposed PC-12 cells (0.1 and 1 µM). The same endpoints were evaluated in the brain of life-long arsenic exposed rats, from gestation through lactation until adult age (3 ppm, drinking water). *Results:* PARP-1 inhibition was demonstrated through decreased immunoreactivity to PAR polymers and increased concentrations of PARP-1 substrate NAD+ in PC12 cells and in regions of the rat brain involved in learning and memory. Also, exposed animals showed poor performance in an associative memory test. *Discussion and conclusion:* The demonstrated PARP-1 inhibition in the nervous system of arsenic exposed rats and in PC-12 cells may be related to arsenic-induced cognitive deficits. Besides alterations of DNA methylation, this inhibition is another epigenetic modification associated with exposure to a neurotoxicant.

Keywords: PARP-1, arsenic toxicity, epigenetic mechanisms, chromatin remodeling

INTRODUCTION

The most prominent non-lethal effect of arsenic after prolonged low-level exposure is the compromise of cognitive abilities of children, revealed as reduced scores in measurements of intellectual function (Rodriguez et al., 2003; Wasserman et al., 2007; Rosado et al., 2007). The mechanisms underlying cognitive deficits have been explored in animal models of chronic exposure to environmentally relevant arsenic concentrations in drinking water, where behavioral and neurochemical parameters have been assessed. An association has been demonstrated between arsenic exposure and decreased performance on different behavioral tasks such as the water maze (Wang et al., 2009), conditioned flavor aversion (Garcia-Medina et al., 2007) and contextual fear conditioning (Martinez et al., 2011). The behavioral impairments observed involve dysfunctions of several brain regions, hippocampus and frontal cortex among them, that could be related to alterations of cell signaling through neurotransmitters and intracellular messengers (Rodriguez et al., 2003; Wang et al., 2010). Furthermore, arsenic exposure modifies the expression of certain genes in human population, and now we know that those changes involve epigenetic mechanism such DNA and histone methylation

(Martinez et al., 2011). These modifications are of crucial importance for the central nervous system since not only during development but also in fully differentiated neurons, epigenetic mechanisms are involved in the continuous formation, reinforcing and elimination of synapses, which constitutes neuronal plasticity.

Among the important enzymes that participate in epigenetic modifications poly (ADP-ribose) polymerase protein 1 (PARP-1) regulates DNA replication, transcription and repair (Lunec et al., 1984; Park et al., 1983) and displays an intimate connection with essential cellular processes such as intracellular signaling, transcription, DNA repair, telomere maintenance, cell cycle and death, among many others. This enzyme is responsible for 90% of poly-ADP modifications in the cell (Bürkle, 2005), and modifies enzyme activity and gene transcription by transferring ADP ribose units from NAD+ on glutamic and aspartic acid carboxylic residues in a variety of nuclear proteins including PARP-1 itself (Burzio et al., 1979; Ogata et al., 1980 a, b; Riquelme et al., 1979).

Several studies have demonstrated the participation of PARP-1 in epigenetic mechanisms of chromatin remodeling regulating transcription of genes involved in learning and memory processes. In the marine mollusk *Aplysia californica* poly (ADP-ribosylation) of histone H1 occurs during long-term facilitation in a test of sensitization (Cohen-Armon et al., 2004). More recently, Fontan-Lozano and collaborators (2010) demonstrated that poly (ADP-ribosylation) of histone H1 is required for synaptic plasticity in the process of memory consolidation. Another research group described additional roles of PARP-1 in processes regulated by the cascade of phosphorylation of mitogen-activated protein kinase, which include growth and differentiation of neurites as well as formation of long-term memory in mammals (Goldberg et al., 2009).

Studies performed in mice designed to evaluate the participation of PARP in the process of memory did not establish a single role for this enzyme, but a series of events that can be interpreted according to the wide spectrum of PARP-1 targets. The partial inhibition of PARP-1 in mice with traumatic brain damage maintains the levels of NAD+ cells, which leads to improvement in the performance of animals in behavioral tests as compared with those animals not receiving the PARP-1 inhibitor (Satchell et al., 2003). This happens because PARP-1 activation induced cell damage depletes cellular NAD+ and consequently adenosine triphosphate (ATP) levels, and this depletion is associated with a cell death that can be prevented by PARP inhibitors. In these cases, since trauma activates a PARP-1 activation that can lead to cell death, the local administration of a PARP-1 inhibitor in the dorsal hippocampus prevents the depletion of NAD + and modestly improves the performance of mice in a test of spatial memory (Clark et al., 2007). On the other hand, complete inhibition of PARP-1 impairs spatial memory acquisition. This is associated with inhibition of ribosylation of proteins implicated in memory and learning, where PARP-1 participates as an epigenetic modulator (Satchell et al., 2003).

The main interest for the present investigation is to explore whether arsenic inhibits PARP-1 in neural cells and in brain regions related to learning and memory. Arsenic inhibits enzymes with zinc finger domains, such PARP-1, due to its high affinity for sulfhydryl groups of cysteines in this domain (Ding, et al., 2009). Since arsenic is involved in processes of carcinogenesis, its effects on cellular signaling pathways in non-neuronal cells have been intensively studied, demonstrating a strong link between arsenic and PARP function. Qin and collaborators (2008) demonstrated that 2 µM arsenite enhances ultraviolet radiation-induced DNA damage in Ha-Cat cells by inhibiting PARP-1 activity, a finding compatible with the known role of arsenic as co-carcinogen. In fact, DNA damage associated with arsenic exposure is dramatically increased in embryonic fibroblasts from PARP-1-null mice, indicating that PARP-1 is required to counteract the cytotoxic effects of arsenic in mammalian cells (Poonepalli et al., 2005).

In light of the observed link between PARP and learning processes, this work was designed to evaluate the *in vivo* and *in vitro* inhibition of PARP-1 by arsenic. Since catecholamine synthesis and release are targets of arsenic (Rodriguez et al., 1998; 2003), PC12 cells represent an appropriate *in vitro* model to study arsenic neurotoxicity because they display a catecholaminergic phenotype. In fact, recent works demonstrate that exposure to low micromolar levels of sodium arsenite for five days result in reduced neurite production, outgrowth and complexity in newly differentiating PC12 cells and Neuro-2a (N2a) neuroblastoma cells (Frankel et al., 2009; Wang et al., 2010). For *in vivo* experiments, on the other hand, we employed a rat model of arsenic exposure from gestation to adult age that is associated with behavioral, neurochemical, morphological and epigenetic alterations (Zarazua et al., 2006, 2010; Ríos et al., 2009; Martinez et al., 2011). In both *in vivo* and *in vitro* studies, NAD+ levels and the presence of poly (ADP-ribosylated) proteins were employed to demonstrate PARP-1 activity.

METHODS

Reagents and Antibodies

Most chemicals were analytical grade reagents and were dissolved in molecular grade water (18.2 Ω/cm). For *in vitro* experiments, PC 12 cells were acquired at ATCC (Manassas, VA); salts, enzymes, sodium arsenite and the protease inhibitor cocktail were obtained from Sigma-Aldrich (St. Louis, MO, USA). Cell culture media (DMEM and Ham F-12) and fetal bovine serum were purchased from GIBCO BRL (Carlsbad, CA, USA), while the BCA protein assay reagents were from Pierce (Thermo Scientific, Rockford, IL, USA). The PARP-1 inhibitors 3-aminobenzamide (3AB) and PJ-34 were provided by Sigma-Aldrich (St. Louis, MO, USA). For immunofluorescence the anti-PAR antibody was obtained from AXXORA (San Diego, CA, USA) and the FITC-coupled anti-mouse secondary antibody from SEROTEC (Raleigh, NC, USA).

The *in vivo* experiments required the same salts and enzymes for NAD quantification used for *in vitro* assays, as well as the PARP inhibitor 3AB. In addition, for the immunostaining of brain regions a biotinylated rabbit anti-mouse secondary antibody was obtained from DAKO (Carpinteria, CA, USA), instead of the fluorescent secondary antibody used for *in vitro* experiments.

In Vitro Experiments

Cell Culture and Experimental Design

PC12 cells were grown in DMEM/F-12 supplemented with 10% fetal bovine serum (FBS), in standard culture conditions (37 °C, 5% CO_2 and 95% humidity). They were plated either on 25 cm^2 culture flasks for NAD^+ determination, or in 24-well plates with poly-L-lysine coated glass coverslips for immunofluorescence, and 50% of medium was replaced every other day with 37° C fresh medium. When 50-60% confluence was reached, differentiation was induced by shifting to differentiation medium, which consisted of DMEM containing 15% FBS and supplemented with 50 ng/ml nerve growth factor (NGF). The medium was completely refreshed every 3 days and differentiation was completed after 14 days and remained stable until day 21. All experiments were performed on day 21 of differentiation. On day 20 of differentiation, the different treatments were added to each flask as follows: differentiation medium as control, sodium arsenite 0.1 and 1 µM, 3-

AB 10 µM and PJ-34 150 nM in differentiation medium. The concentrations of inhibitors were based on the IC_{50} values reported in the literature (Iwashita et al., 2005). Sodium arsenite stocks were prepared in double-distilled water and sterilized using a 0.22-µm syringe filter. Working solutions were prepared by diluting the stock with differentiation medium. After 24 h treatment the cells were either homogenized for NAD+ determination or fixed for immunofluorescence analysis.

NAD⁺ Determination

For determination of NAD+ levels, the cells were separated using a scraper, collected in Eppendorf tubes and centrifuged at 1100 g 4 ° C for 5 minutes. The reaction mixture contained glycylglycine buffer (0.065 M glycylglycine, 0.1 M nicotinamide, 0.5 M ethanol, pH 7.4), alcohol dehydrogenase, thiazolyl blue [bromide 3-(4.5 - dimethyl-azal-2yl)-2,5 diphenyl tetrazolium (MTT)] and phcnazinc mcthosulfatc (PMS). Thc rcaction was allowed to run for 7 minutes and the absorbance determined at 556 nm (Clark et al., 2007; Nisselbaum and Green, 1969; Satchell et al., 2003). NAD + levels in the cell samples were calculated as µmol/µg of protein, the proteins were determined by absorbance at 280 nm. Through a standard validation procedure, this technique showed linearity ($r^2 = 0.99$) and 100.9 % recovery, with coefficients of variation ranging between 3.2 and 3.6 % for repeatability and intra-laboratory reproducibility, respectively.

Immunofluorescence for Poly (ADP-Ribosylated) Proteins

On day 21 after differentiation, the cells cultured in poly-L-lysine coated glass coverslips were exposed to different treatments (control, arsenic, 3-AB and PJ-34) were washed with cold PBS (1X, pH 7.4) and fixed with cold 4% paraformaldehyde for 15 minutes, followed by three successive washes of 10 minutes with PBS and stored at -20° C in cryoprotectant solution (30% glycerol, 30% polyethylene glycol and 40% PBS 0.05M) until processed for immunofluoresence. To label proteins with fluorescence the coverslips were washed with PBS, incubated in HCl 2N, at 37°C for 30 minutes, washed twice with boric acid 0.1 M, rinsed with PBS, followed by $NaBH_4$ 0.05% during 15 minutes with the purpose of decreasing the background. Then, the coverslips were incubated in blocking reagent (10% AB serum plus 0.25% Triton in PBS) at room temperature for 2 h, followed by overnight exposure to a mouse monoclonal antibody against poly (ADP -ribose) (PAR) diluted 1:300 in a mixture of 1% AB-serum plus 0.25% Triton X-100^{TM} in PBS, at 4 ° C. Afterwards, the coverslips were incubated for 2 h in the darkness with the anti-mouse secondary antibody conjugated with FITC (diluted 1:300 in the same solution) followed by counterstaining with propidium iodide to visualize the nucleus. Between each incubation mixture the coverslips were washed three times for ten min each with 1% AB serum, Triton X-100^{TM} 0.25% in PBS. Finally, the coverslips were mounted on slides using Mowiol containing 2.5% DABCO^{TM}. The coverslips were processed in batches containing all experimental groups in order to avoid differences due to sample handling. The images were obtained using a fluorescence microscope (Leica DM2500) at 100X magnification. Using a digital camera (Leica DFC 345FX) coupled to the microscope seven photographs per coverslip were taken (4 slides containing 2 coverslips for each treatment) and analyzed using Adobe PhotoShop C3 (Adobe Systems Incorporated, San Jose, CA, USA) for counting red stained nuclei and positively stained cells for PAR, which were observed as green fluorescent dots. Image merging allowed verifying which cells contained PAR positive nucleus.

In Vivo Experiments

Animal Model and Experimental Design

The experiments were performed in our previously reported animal model of life-long exposure to arsenic (Martinez et al., 2011; Rios et al., 2009; Zarazua et al., 2010). This experimental design was performed according to the "Guidelines for the Care and Use of Mammals in Neuroscience and Behavioral Research" (2003) and the protocol was approved by the ethics committee of Universidad Autonoma de San Luis Potosi. Briefly, 24 female and 24 male Wistar rats weighting between 200 and 250 g were assigned to three experimental groups of eight couples each. Two groups received arsenic-free drinking water: one for raising the litters serving as controls treated with vehicle only, and the other for the animals that were to be treated with 3-AB, (30 mg/kg i.p. during seven days). The third group had access to water containing 3 ppm of sodium arsenite, a concentration that resulted in an approximate ingestion of 0.3–0.4 mg/kg/day of arsenic, equivalent to an exposure level that has already been reported in human populations (Hsieh et al., 2008). Each mating pair was placed in a cage and maintained under a 12 h light/dark cycle with food (Prolab RMH2500, PMI Nutrition International, Brentwood, MO) and water *ad libitum*. After ten days the males were removed and arsenic exposure to the female rats continued throughout gestation and lactation periods. Offspring were weaned at 4 weeks, separated by sex and subjected to continued arsenic exposure for 4 months. The experiments were performed in groups of rats belonging to different litters for each treatment as follows: (a) groups of three males for immunohistochemical studies, (b) groups of six males which were used first for the behavioral tests followed by NAD+ quantification in their brains. As previously reported, monthly recording of body weight confirmed no significant difference between control and arsenic exposed animals (Martinez et al., 2011)

NAD+ QUANTIFICATION

Animals were sacrificed by decapitation 24 hours after completing the behavioral tests, the brain removed from the skull, and the regions of prefrontal cortex and hippocampus bilaterally dissected out over ice. Tissue samples were immediately homogenized in KPO_4 buffer (0.05 M, pH 6.0) containing 0.1 M nicotinamide, keeping them on ice at all times. Afterwards the homogenates were heated in boiling water for 5 min, and then chilled out on ice for another 5 min. The samples were centrifuged at 15000 g for 10 minutes at 4 ° C. The supernatants were stored at -70 ° C until analysis as a group, according to the procedure described above for NAD+ determination in cell cultures. NAD+ levels in the samples were calculated as µmol/mg wet weight. The validation parameters in homogenates were as follows, repeatability % coefficient of variation = 4.1, reproducibility % coefficient of variation = 4.3, linearity r^2 = 0.986 and % recovery = 99.6 %.

Immunohistochemistry for Poly (ADP-Ribosylated) Proteins

At the time of sacrifice, three rats from each experimental group were deeply anesthetized with sodium pentobarbital (50 mg/kg, i.p.), and perfused by gravity through the ascending aorta with 300 mL of 0.1 M PBS (pH7.4) followed by 300 mL of 4% paraformaldehyde and 0.6% glutaraldehyde in 0.1 M phosphate buffer (pH7.4). Brains were removed, the hemispheres separated from the cerebellum and blocked into rostral and caudal sections, through a coronal cut approximately 1 mm posterior to bregma (Paxinos and Watson, 2005). Tissue blocks were immersed in 0.1 M phosphate buffer containing 4% paraformaldehyde and 0.6% glutaraldehyde, during 24h at 4 ° C, followed by embedding in paraffin. Coronal brain sections (7 µm) made at the level of the frontal cortex (between 3.2 and 3.7 mm anterior to bregma), and the region of the dorsal hippocampus (located at about -2.1 mm from bregma) were collected in silanized slides. Next, sections were dewaxed, rehydrated and subjected to a sequence of incubation steps in a humidity chamber, starting with sodium citrate buffer (0.1 M, pH 6) at 100°C during 30 min for epitope recovery (Lorincz and Nusser, 2008), followed by 2 N HCl during 15 min at room temperature with the purpose of exposing poly(ADP-ribosylated) proteins, and finally with 3% hydrogen peroxide in methanol during 15 minutes for blocking endogenous peroxidase. Sections were incubated overnight at 4 °C with 1:50 mouse monoclonal anti-poly (ADP-ribose) antibody followed by the streptavidin–biotin marked secondary antibody for 15 min at room temperature. Peroxidase activity was visualized by incubating the sections with diaminobenzidine, and the sections were counterstained with Harris hematoxiline. All rinses between the incubation steps were performed with TBST (Tris buffer saline–tween, 0.05 M, pH 7.4). The sections were processed in batches that included 2 slides with 3 sections each from each treatment (control, arsenic, and 3-AB),. A negative control was included which consisted of tissue sections treated without the primary antibody for each batch. The sections were viewed with a Nikon microscope (Nikon, Labophot-2, Japan) equipped with a digital camera. Three photographs per section for hippocampus and four for cortex (2 slides containing 3 sections for each region) were taken (3 images for hippocampus and 4 images for cortex), and analyzed for staining intensity using the software Image J 1.44 k (National Institutes of Health, USA). All color images obtained at the same magnification (40X) were digitally transformed to a grey scale, and the intensity of each image was assessed by applying a uniform digital filter to all of them. Approximately 18 values for hippocampus and 24 for cortex obtained from each rat were averaged to yield a final measurement of immunostaining intensity per animal. The quantitative results are given in pixels, and they arise from 3 animals for each treatment (control, arsenic, and 3-AB).

Behavioral Testing

Open Field Test

The open field was a square arena (100 cm × 100 cm; height: 50 cm) constructed with expanded polystyrene. The maze was equally divided into 25 quadrants and located in a room with dim lighting. Rats of each experimental group were evaluated in a random order. After placing the animal in the center of the arena, its movements were observed during ten minutes by one person who was blind to the treatment of each rat. The arena was cleaned thoroughly with 1% acetic acid after completion of the trial. Open field measures of general activity included the total number of quadrant crossings and rearings on the center and periphery. The total number of crossings was determined by how often the four paws of an animal entered a quadrant from an adjacent quadrant, rearings were noted when a rat stood on its two hind paws. Also the latency to find the lateral borders of the arena and the number of droppings were recorded as indicators of emotional reactivity (Sahgal, 1993).

Morris Water Maze

Spatial memory acquisition of rats in the Morris water maze (MWM) was assessed by observers unaware of the experimental group, as previously described (Castillo, 2002). Briefly, a white pool (140 cm diameter, 50 cm deep) filled with 25°C opaque water to 30 cm depth was situated in a room with several extra maze cues located on the walls. About 1 cm below the surface of the water was hidden a 10 cm round goal platform, at approximately 40 cm from the pool wall. Testing was performed during seven days when animals were 15 weeks old. Each rat was subjected to a series of four trials/day. For each trial, rats were randomized to one of four starting locations and placed in the pool. Rats were given a maximum of 300 s to find the submerged platform. When they succeeded in finding the platform they were allowed to remain there for 30 s between trials, otherwise the animals were manually placed in the platform and left there for one minute. Performance in the Morris-water maze was quantified by measuring the latency to find the platform, and counting the number of crossings, defined as every time when the animal left the periphery swimming towards the center of the pool.

Contextual Fear Conditioning

The training chamber (acrylic, 20 cm x 25 cm x 25 cm) for the contextual fear conditioning (CFC) test was located inside a larger wooden box in order to isolate the chamber from the surrounding environment and to allow only one person to observe the rats. The observer was blind to the treatment to which each animal had been subjected to. Training started by placing the animal inside the chamber and allowing an exploration period of 2 min after which the rat received the conditioning stimulus (CS) consisting of a four second burst of a 4 dB tone followed by 2 second electric shock (US) of 0.75 mA. Three pairs of CS/US were applied in 30 seconds intervals after which animals were given 30 additional seconds in the chamber before returning them to their cage. The evaluations of each animal were performed after 1, 6, 24 and 96 h of the conditioning phase. After placing the animal in the cage and allowing a 30 s exploration period, the CS (tone) was applied resulting in a characteristic immobility behavior (Colon et al., 2006), which was measured for a maximum of 5 min before returning the animal to its cage. The comparison between total freezing times indicated the difference between treatments at each testing point, while the analysis of the response over time for each group allowed the evaluation of the extinction of the conditioned response.

Statistic Analysis

Normality data distribution and variances homogeneity were evaluated by Shapiro Wilk and Levene tests, respectively. For the data of NAD+ levels in cells and brain, the groups were either compared by means of ANOVA followed by Tukey-Kramer multiple comparison tests or through Student's t-test for comparisons of two groups. The data of CFC and MWM tests were analyzed with a two-factor, repeated measures ANOVA (time and treatment as factors), followed by a Bonferroni post test. The results of immunofluorescence were analyzed by a X^2 test with Bonferroni correction in order to compare experimental groups against the controls, considering the number of positive cells as dichotomous variables. The results of immunohistochemistry were analyzed by means of ANOVA and Tukey-Kramer post test when the data were normally distributed, and Kruskall Wallis test followed by X^2 when they did not show a normal distribution. These statistical analyses were performed with two commercial software packages (Statistica 6.0, and GraphPad Instat 3). Values of p below 0.05 were considered statistically significant in all cases.

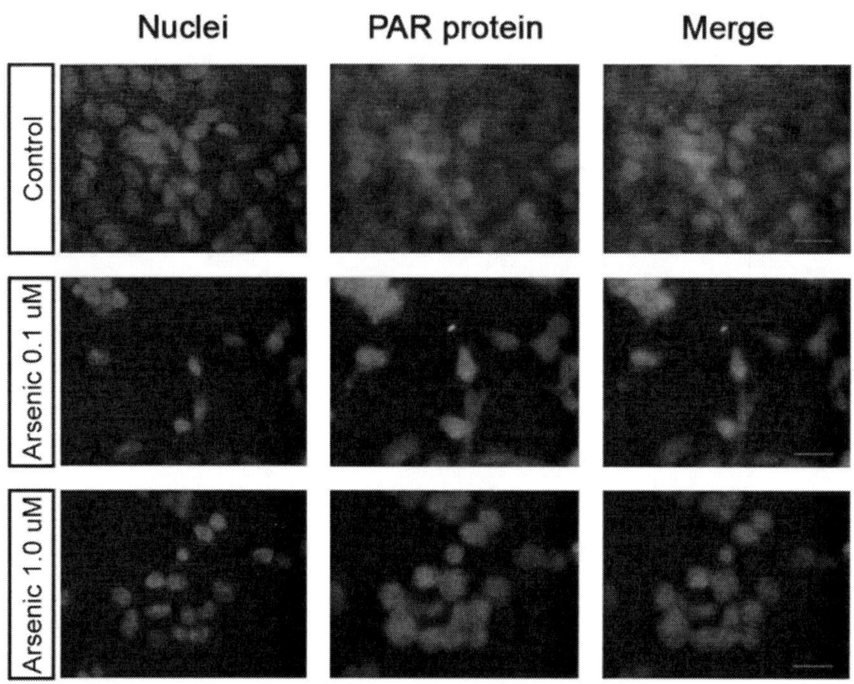

Figure 1. Control cells show a basal poly (ADP-ribosylation), characterized by the fluorescence of nuclear proteins that are seen as green dots in the column named PAR, and the merge image confirms their nuclear localization. Cells treated with arsenic 0.1 µM and 1 µM show a clear, concentration-dependent decrease of poly (ADP-ribosylation). Cells were labeled with FITC for poly (ADP-ribosylated) proteins and propidium iodide for nuclei. Scale bar equals 10 µm.

RESULTS

Arsenic Inhibits Poly-ADP-Ribosylation in PC-12 Cells

After complete NGF-induced differentiation of PC-12 defined by the presence of two or more neurites with a length equal to or greater than twice the diameter of the cell body, two set of experiments were performed in order to determine whether arsenic inhibits PARP-1 activation in these cells. First, poly-ADP-ribosylation assessed by immunofluorescence revealed a basal pattern of poly-ADP-ribosylation in control cells (Fig 1 A), which was manifested as dots of strong fluorescence intensity inside the nuclei, while 24-h arsenic exposure resulted in a dose-dependent reduction of poly-ADP ribosylation (Fig 1 B and C). Fig 2 shows that cells treated with specific PARP inhibitors had a significant reduction of poly-ADP ribosylation, whose numerical results are shown in Table 1 (all groups analyzed with Bonferroni test, p < 0.012 vs. vehicle).

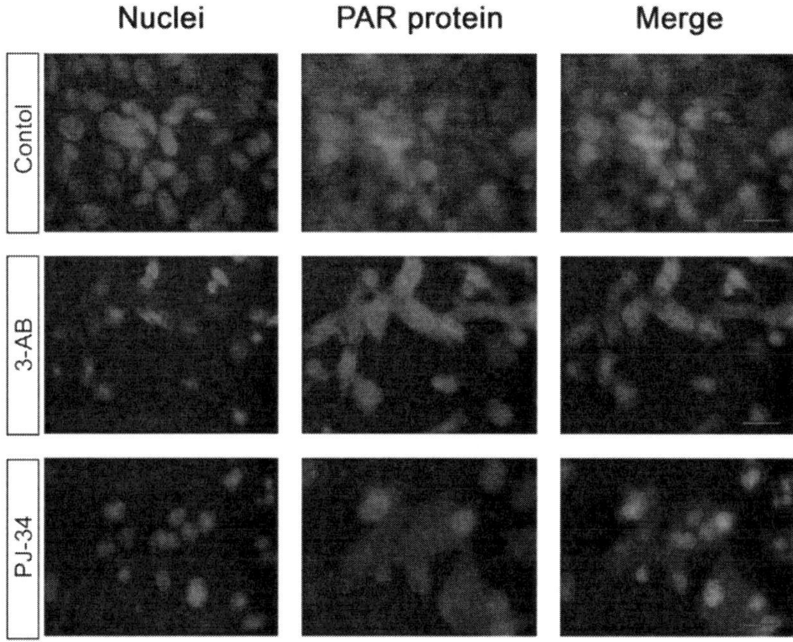

Figure 2. As compared with control cells a decreased poly (ADP-ribosylation) is observed in cells treated with 10 μM 3-AB and cells treated with 150 nM PJ-34. Cells were labeled with FITC for poly (ADP-ribosylated) proteins and propidium iodide for nuclei. Scale bar equals 10 μm.

Table 1. PC-12 cells labeled for poly (ADP-ribosylated) proteins

	Total cells	Negative cells	Positive cells	% positive cells
Control	1160	1106	54	4.65
As 0.1 uM	1421	1397	24	1.68 *
As 0.1 uM	1724	1710	14	0.81 *
3-AB 10 uM	1391	1366	25	1.83 *
PJ-34 150 nM	1533	1523	10	0.65 *

Counts of negative and positive poly (ADP)-ribose) labeled cells. All cells were counterstained with propidium iodide to visualize the nucleus, which represents the number of total cells in the 7 fields analyzed. Results are from 3 independent experiments. * Significant difference from control group, X^2 test with Bonferroni correction.

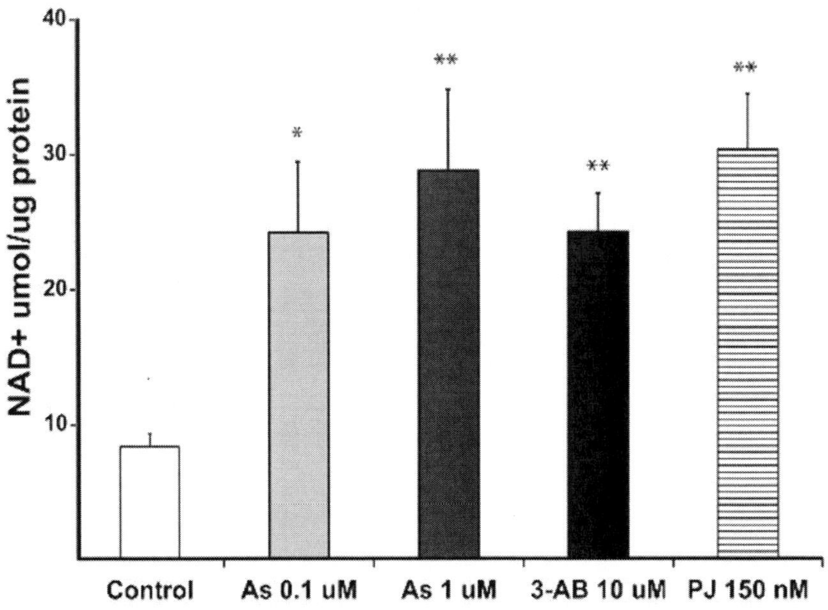

Figure 3. Treatment with arsenic in two different concentrations or with inhibitors of PARP-1 is associated with an increase of NAD+ concentration, indicating a decreased substrate utilization for the production of ADP-ribose. Asterisks represent significant differences between groups analyzed by ANOVA followed byTukey-Kramer post-test. * p< 0.050 vs. control. ** p< 0.010 vs. control. Values represent the mean ± SEM of 8 independent experiments.

Next, quantification of NAD+ levels in PC-12 cells after 24 h exposure to 0.1 or 1 µM arsenic resulted in 189 % and 243 % NAD+ increases, respectively, although this change was only significant for the larger concentration tested (As 0.1 µM vs. control, p > 0.050; As 1 µM vs. control, p < 0.050). When NAD+ levels were examined in cells treated with 3-AB and PJ-34 increases of 190 % and 262 % respectively were obtained, but only PJ-34 showed a significant increase (Fig 3. p < 0.010 vs. vehicle).

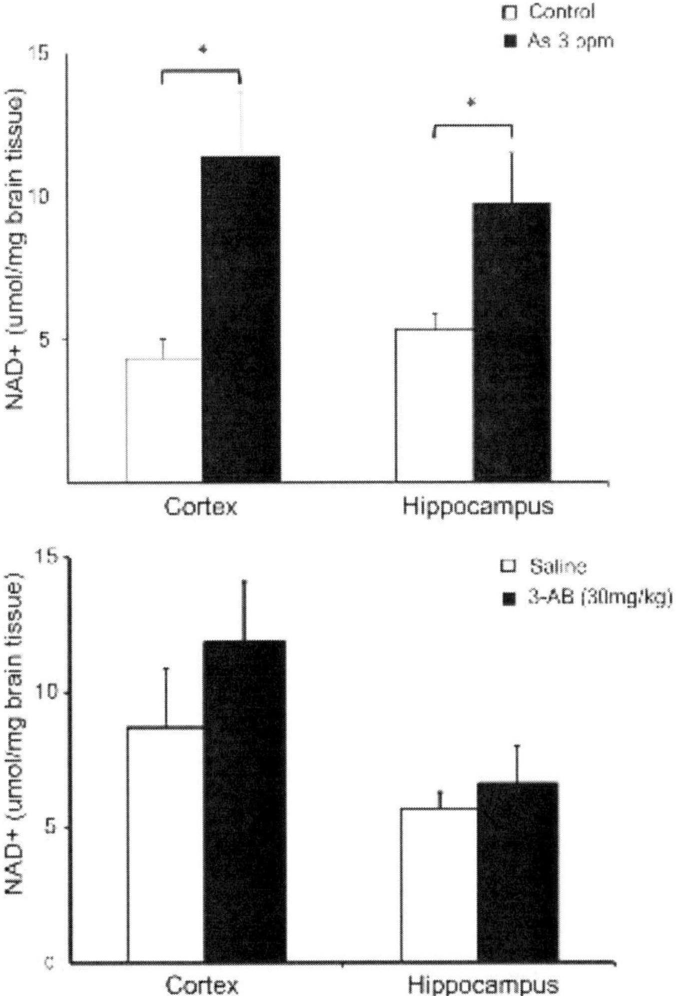

Figure 4. Measurements of NAD+ concentration in rat cortex and hippocampus, expressed as μM/mg brain tissue. (A) Chronic arsenic exposure is associated with a significant increase of NAD+ concentration in cortex and hippocampus, bars represent means ±SEM obtained from 8 animals per group, * p < 0.050, Student´s t-test for both brain areas. B) 3-AB treatment produces no changes in NAD+ concentration in cortex and hippocampus, bars represent means ±SEM obtained from 6 animals per group, * p > 0.050, Student t-test for both brain areas.

Chronic *In Vivo* Arsenic Exposure Increased NAD+ Levels in Hippocampus and Frontal Cortex of Rats

The larger increase of NAD+ levels associated with arsenic exposure was observed in cortex, where the control group had a concentration of 4.3 ± 1.9 µM/mg brain tissue while in the arsenic exposed group it was 11.3 ± 6.6 µM/mg (p = 0.003). A similar effect, albeit of lower magnitude, was observed in the hippocampus, where arsenic caused a significant increase in the NAD+ levels (9.7 ± 5.0 µM/mg) in comparison with the control group (5.3 ± 1.6 µM/mg; p = 0.047; Fig 4 A). However, in the rats treated with the specific PARP inhibitor 3-AB no significant changes were observed neither in cortex nor hippocampus (Fig 4 B; p= 0.337 and p= 0.567 vs. vehicle, respectively).

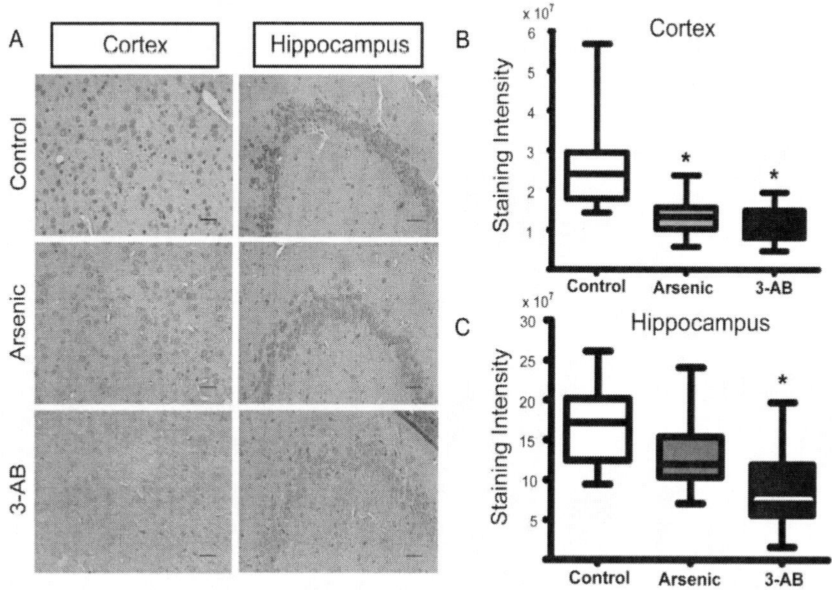

Figure 5. Left panel. Photomicrographs of 7 µm coronal sections showing poly (ADP-ribose) positive nuclei in, control, arsenic and 3-AB for cortex and hippocampus (A). Scale bar equals 100 µm. Right panel. Poly (ADP-ribosylation) expressed as staining intensity for cortex B, both arsenic and 3-AB groups are different from control by X^2 analysis, * p < 0.001, and for hippocampus (C) only 3-AB group are different of control by Tukey analysis, * p < 0.001.

Arsenic Exposure is Associated with Decreased Poly (ADP-Ribosylated) Proteins in Rat Frontal Cortex

The immunoreactivity to poly (ADP-ribosylated) proteins was assessed in situ in sections of hippocampus and frontal cortex. As shown in Fig 5, significant effects of treatments were found in both regions. The control group presented basal levels of immunostaining in the cortex, followed by a significant decrease of poly (ADP-ribosylation) in the arsenic and 3-AB treated groups (Fig 5 D; p < 0.001 vs. control in both cases). In contrast, non significant changes of immunoreactivity were observed in the hippocampus of arsenic-exposed rats (p = 0.072 vs. control), indicating a lack of inhibition of PARP-1 in comparison with the control group, which contrasted with the significant inhibition observed in the 3-AB-treated rats (Fig 5 E; p = 0.0001 vs. control).

Effect of PARP-1 Inhibition on Motor Function

At 15 weeks of age all rats were tested in the open field to determine whether arsenic or 3-AB induced gross alterations of motor function. The experimental results revealed no significant effects between arsenic-exposed rats or treated with 3-AB (30 mg/kg) vs. the control group in any variable assessed, such as latency to find the edge, locomotion, rearings, or number of droppings (Student´s t-test, p > 0.5 in both cases; data not shown).

Contextual Memory Deficits Induced by Arsenic or PARP-1 Inhibitor Exposure

The animals were tested at 15 weeks of age in order to sacrifice them when reaching four months of age and make determinations of NAD+ and poly (ADP- ribosylation). In the fear conditioning test, arsenic-exposed rats showed significantly shorter freezing periods at 6, 24 and 96 h after training in comparison with their controls (Fig 6 A; p = 0.043). Although a tendency to decrease freezing times was observed in animals treated with 3-AB, the changes were not significantly different from vehicle-treated animals (Fig 6 B; p =0.068). Extinction was a second parameter in which groups of animals were analyzed. All three groups of animals showed a significantly lower freezing time 96 h after training.

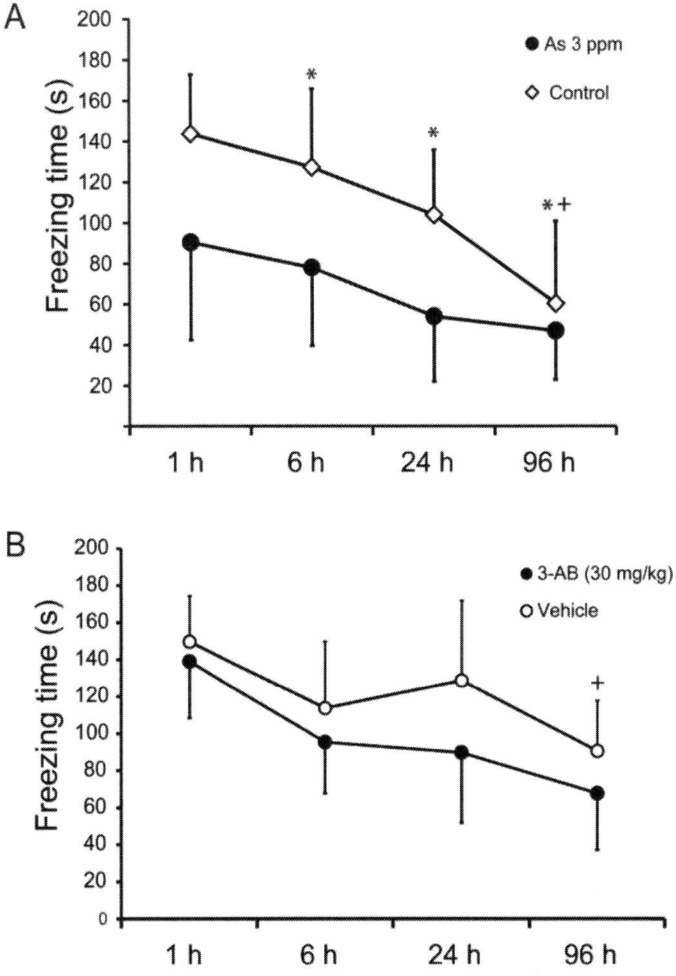

Figure 6. Contextual Fear Conditioning, freezing time (s) presented by each group of animals 1, 6, 24 and 96 h after conditioning. (A) Arsenic group had significant shorter immobility times in comparison with controls. Two-factor repeated measures ANOVA followed by Bonferroni posttest: * p = 0.043 for treatment, and † p < 0.001 for time (96 h vs. 1 h in both groups). Points represent means ± SEM of 8 animals/group. (B) No significant differences were found for immobility times between 3-AB group and controls. Two-factor repeated measures ANOVA followed by Bonferroni posttest: p = 0.068, and † p < 0.001 for time (96 h vs. 1 h in both groups). Points represent means ± SEM of 6 animals/group.

Spatial Memory in the Morris Water Maze

The control and arsenic exposed animals showed a similar spatial memory acquisition manifested as a typical pattern of decreasing latency to find the platform over the seven testing days, without significant differences between groups, (p = 0.978 for treatment, p < 0.0001 for time). Also in the animals treated with 3-AB, no significant differences in latencies to find the hidden platform were observed (p = 0.061 for treatment, p < 0.0001 for time). There were no differences in the number of crossings within control and treated groups, suggesting that differences in performance were not related to persistent motor function deficits (data not shown).

DISCUSSION

This study demonstrates PARP-1 inhibition in neural cells exposed *in vitro* to arsenite. Also, rats exposed throughout their whole lives to arsenite presented PARP inhibition in the hippocampus and cerebral cortex, which are brain regions critically involved in learning and memory processes.

The results obtained in cell cultures indicate that exposure to arsenite at environmentally relevant concentrations such as 0.1 μM was associated with an inhibition of PARP-1 comparable to that caused by the PARP-1 inhibitor 3-AB, while exposure to a 10-fold higher arsenite concentration causes an inhibition comparable to that produced by a more specific and potent inhibitor PJ-34 (Fig 1 and 2). In addition to the decrease of poly (ADP-ribosylation) of nuclear proteins in PC-12 cells, we also observed a significant increase of NAD+ in cells exposed to arsenite and to PJ-34, which is consistent with the inhibition of PARP-1 since the polymers of ADP-ribose are synthesized from NAD+ (Bürkle, 2005). It is important to note that poly (ADP-ribosylation) was inhibited in absence of DNA damage, and that trivalent methylated arsenicals are also able to inhibit this basal PARP-1 activity in non-neural human cells in culture (Walter et al., 2007). Thus, besides its role as co-carcinogen through inhibition of DNA repair (Qin, 2008) other events related to PARP-1 activity are disturbed by arsenite. These would include PARP-1 acting as a downstream target of phospholipase C, where inositol trisphosphate-Ca^{2+} mobilization directly induces a fast poly ADP -ribosylation (Homburg et al., 2000), its participation in neurotrophic NGF effects in PC-12 cells (Visocheck et al., 2005) and PARP-1 activation induced by ERK2 which promotes growth, proliferation and differentiation regulated by the Raf–MEK–ERK phosphorylation cascade (Cohen-Armor et al., 2007). All previous reports of PARP-1 activation in catecholaminergic cells refer to its involvement in energy depletion and cell death when a neurotoxic substance such as MPTP or 6-OHDA induce cell damage. As in non neural cell types, PARP-1 activation contributes to cell death (Cossi and Marien, 1999; Lee et al., 2007). In those cases PARP-1 inhibition would be a useful target to prevent neuronal death. No data linking PARP-1 inhibition with changes of catecholaminergic parameters that could be deleterious for synaptic plasticity have been generated until now.

In vivo, decreased PARP-1 activity in the brain of arsenite exposed animals was demonstrated by significantly increased NAD+ concentrations in both the cortex and the hippocampus, as compared with non exposed animals that underwent the same behavioral testing (Fig 4 A). Also, poly (ADP-ribosylation) of nuclear proteins was significantly reduced in cortical cells of arsenite exposed rats (Figure 5). These results are relevant despite the fact that the rat is not the best model for arsenic toxicity. The protein sequence of rat hemoglobin results in arsenic binding to the protein and storage in red blood cells, while human and mice hemoglobin do not present this characteristic (Lu et al., 2007). Then, in order to study SNC functions under arsenic exposure, the rat model has been validated in terms of producing similar effects as those observed in human exposure (behavioral deficits) even when rats require higher arsenic intake. Rat models are nowadays widely used for *in vivo* experiments that could not be easily performed in mice, such microdialysis (Zarazua et al., 2006) and hippocampal recording (Kruger et al., 2007) among others.

Concerning the inhibition of PARP-1 in animals treated with the PARP-1 inhibitor 3-AB, we observed a significant decrease of immunostaining for poly (ADP-ribosylation) in both regions without changes of NAD+ levels (Fig 4B and 5). We interpret these apparently contradictory data in light of a difference between our experimental design and those reported in the literature, where the effects of PARP-1 inhibitors have been demonstrated after PARP-1 stimulation (Abdelkarim et al., 2001; Besson et al., 2005; Iwashita et al., 2005 Fontan-Lozano et al., 2010). In our study we only measured basal NAD+ levels with and without arsenic, finding that the inhibition of basal PARP-1 activity does not necessary induce a significant increase of NAD+, due to the multiple roles of this nucleotide. In chronic arsenic exposure, however, we observed arsenite inhibition of basal PARP-1 activity, leading even to increased NAD+ concentration, possibly due to the long term duration of the phenomenon.

Finally, this animal model of arsenic exposure demonstrated that four month aged animals exposed to 0.3–0.4 mg/kg/day of arsenite display poor performance in an associative memory test (CFC, Fig 6), as previously reported (Martinez et al., 2011). Although arsenic is known by its effects in sensory peripheral nerves, the doses used in this study is very low as compared to those that induce peripheral nerve damage and affect conduction velocity (García-Chavez et al., 2007). Also in humans, high exposure levels are not associated with alterations of nerve conduction velocities in every population tested (Fujino et al., 2006; Tseng et al., 2006) In contrast, no significant differences were found in spatial memory (MWM), possibly because arsenite exposure was below the level needed to impair performance in this test. In support of this possibility, deficits in the MWM have been associated with arsenic doses more than 10 times higher than those used in this study (5.4 -7.7 mg/kg/day, Luo et al., 2009).

CONCLUSION

Arsenic is a potent cellular toxin which increases ROS production, induces DNA damage and energy depletion that can eventually lead to cell death. However, low levels of exposure exert remarkable CNS effects which often lead to cognitive deficits. The molecular events that underlie such effects have been difficult to trace. Until now, previous studies have demonstrated methylation anomalies in genes involved in synaptic plasticity (Martinez, 2011), decrease of nitric oxide production (Zarazua, 2006), intracellular calcium disturbances (Florea, 2005) and inhibition of long term potentiation (Kruger, 2007). Together with PARP-1 inhibition, all reported arsenic affects can contribute to learning deficits; however the causal role of arsenic on these distinctive events related to memory and learning requires further research. It is the author's opinion that only novel approaches aimed at resolving differences in local epigenetic activity related to synaptic plasticity will provide a better perspective of arsenic-induced disturbances.

ACKNOWLEDGEMENTS

The authors thank D. Cruz and I. Zarazua for technical support, C. Galaviz-Hernández from CIIDIR-IPN Unidad Durango for facilitating research equipment, and to Dr. J. L. Góngora-Alfaro and C. García-Sepúlveda for his review and critics of the manuscript. This work was supported by funding from CONACYT 105937 and a fellowship to F. Ceballos 169086.

REFERENCES

Abdelkarim GE, Gertz K, Harms C, Katchanov J, Dirnagl U, Szabo C, Endres M. Protective effects of PJ34, a novel, potent inhibitor of poly(ADP-ribose) polymerase (PARP) in in vitro and in vivo models of stroke. *Int J Mol Med.* 2001; 7: 255-260.

Besson VC, Zsengeller Z, Plotkine M, Szabo C, Marchand-Verrecchia C. Beneficial effects of PJ34 and INO-1001, two novel water-soluble poly(ADP-ribose) polymerase inhibitors, on the consequences of traumatic brain injury in rat. *Brain Res* 2005;1041: 149-156.

Bürkle A. Poly(ADP-ribose). The most elaborate metabolite of NAD+. *FEBS J* 2005; 272:4576-89.

Burzio LO, Riquelme PT, Koide SS. ADP ribosylation of rat liver nucleosomal core histones. *J Biol Chem* 2005; 254: 3029-3037.

Clark RS, Vagni VA, Nathaniel PD, Jenkins LW, Dixon CE, Szabo C. Local administration of the poly(ADP-ribose) polymerase inhibitor INO-1001 prevents NAD+ depletion and improves water maze performance after traumatic brain injury in mice. *J Neurotrauma* 2007; 24: 1399-1405.

Castillo CG, Montante M, Dufour L, Martínez ML, Jimenez-Capdeville ME. Behavioral effects of exposure to endosulfan and methyl parathion in rats. *Neurotoxicol Teratol* 2002; 24: 797-804.

Cohen-Armon M, Visochek L, Katzoff A, Levitan D, Susswein AJ, Klein R, Valbrun M, Schwartz JH. Long-term memory requires polyADP-ribosylation. *Science* 2004; 304: 1820-1822.

Cohen-Armon M, Visochek L, Rozensal D, Kalal A, Geistrikh I, Klein R, Bendetz-Nezer S, Yao Z, Seger R. DNA-Independent PARP-1 Activation by Phosphorylated ERK2 Increases Elk1 Activity: A Link to Histone Acetylation. *Mol Cell* 2007; 25: 297–308.

Colon CM, Jianpeng W, Ramos X, Garcia HG, Davila JJ, Laguna J, Rosado C, Peña OS. An inhibitor of DNA recombination blocks memory consolidation, but not reconsolidation in context fear conditioning. *J Neurosci,* 2006; 26: 5524-5533.

Cosi C, Marien M. Implication of poly (ADP-ribose) polymerase (PARP) in neurodegeneration and brain energy metabolism. Decreases in mouse brain NAD+ and ATP caused by MPTP are prevented by the PARP inhibitor benzamide. *Ann N Y Acad Sci* 1999; 890: 227-239.

Ding W, Liu W, Cooper KL, Qin XJ, de Souza Bergo PL, Hudson LG, Liu KJ. Inhibition of poly(ADP-ribose) polymerase-1 by arsenite interferes with repair of oxidative DNA damage. J Biol Chem 2009; 284: 6809-6817.

Florea AM, Yamoah EN, Dopp E. Intracellular calcium disturbances induced by arsenic and its methylated derivatives in relation to genomic damage and apoptosis induction. *Environ Health Perspect* 2005; 113: 659-664.

Fontan-Lozano A, Suarez-Pereira I, Horrillo A, del-Pozo-Martin Y, Hmadcha A, Carrion AM. Histone H1 poly[ADP]-ribosylation regulates the chromatin alterations required for learning consolidation. *J Neurosci* 2010; 30: 13305-13313.

Frankel S, Concannon J, Brusky K, Pietrowicz E, Giorgianni S, Thompson WD, Currie DA. Arsenic exposure disrupts neurite growth and complexity in vitro. *Neurotoxicol* 2009; 30: 529-537.

Fujino Y, Guo X, Shirane K, Liu J, Wu K, Miyatake M, Tanabe K, Kusuda T, Yoshimura T. Arsenic in drinking water and peripheral nerve conduction velocity among residents of a chronically arsenic-affected area in Inner Mongolia.*J. Epidemiol* 2006;16: 207-213

García-Chávez E, Segura B, Merchant H, Jiménez I, Del Razo LM. Functional and morphological effects of repeated sodium arsenite exposure on rat peripheral sensory nerves. *J Neurol Sci* 2007; 258: 104-110.

Garcia-Medina NE, Jimenez-Capdeville ME, Ciucci M, Martinez LM, Delgado JM, Horn CC. Conditioned flavor aversion and brain Fos expression following exposure to arsenic. *Toxicol* 2007; 235: 73-82.

Goldberg S, Visochek L, Giladi E, Gozes I, Cohen-Armon M. PolyADP-ribosylation is required for long-term memory formation in mammals. *J Neurochem* 2009; 111: 72-79.

Guidelines for the care and use of mammals in neuroscience and behavioral research. *National Research Council of the National Academies.* The National Academies Press, Washington DC. 2003.

Homburg S, Visochek L, Moran N, Dantzer F, Priel E, Asculai E, Schwartz D, Rotter V, Dekel N, Cohen-Armon M. A fast signal-induced activation of poly(ADP-ribose) polymerase: a novel downstream target of phospholipase C. *J Cell Biol* 2000; 150: 293–307.

Hsieh FI, Hwang TS, Hsieh YC, Lo HC, Su CT, Hsu HS, Chiou HY, Chen CJ. Risk of erectile dysfunction induced by arsenic exposure through well water consumption in Taiwan. *Environ Health Perspect*, 2008; 116: 532-536.

Iwashita A, Tojo N, Matsuura S, Yamazaki S, Kamijo K, Ishida J, Yamamoto H, Hattori K, Matsuoka N, Mutoh S. A novel and potent poly(ADP-ribose) polymerase-1 inhibitor, FR247304 (5-chloro-2-[3-(4-phenyl-3,6-dihydro-1(2H)-pyridinyl)propyl]-4(3H)-quinazolinone), attenuates neuronal damage in in vitro and in vivo models of cerebral ischemia. *J Pharmacol Exp Ther* 2004; 310: 425-436.

Kruger K, Repges H, Hippler J, Hartmann LM, Hirner AV, Straub H, Binding N, Musshoff U. Effects of dimethylarsinic and dimethylarsinous acid on evoked synaptic potentials in hippocampal slices of young and adult rats. *Toxicol Appl Pharmacol* 2007; 225: 40-46.

Lee DC, Womble TA, Mason CW, Jackson IM, Lamango NS, Severs WB, Palm DE. 6-Hydroxydopamine induces cystatin C-mediated cysteine protcasc supprcssion and cathcpsin D activation. *Neurochem Int.* 2006; 50: 607-618

Lorincz A, Nusser Z. Specificity of immunoreactions: the importance of testing specificity in each method. *J Neurosci* 2008; 28: 9083-9086.

Lu M, Wang H, Li XF, Arnold LL, Cohen SM, Le XC. Binding of dimethylarsinous acid to cys-13alpha of rat hemoglobin is responsible for the retention of arsenic in rat blood. *Chem Res Toxicol* 2007; 20: 27-37.

Lunec J, George AM, Hedges M, Cramp WA, Whish WJ, Hunt B. Post-irradiation sensitization with the ADP-ribosyltransferase inhibitor 3-acetamidobenzamide. *Br J Cancer Suppl* 1984; 6: 19-25.

Luo JH, Qiu ZQ, Shu WQ, Zhang YY, Zhang L, Chen JA. Effects of arsenic exposure from drinking water on spatial memory, ultra-structures and NMDAR gene expression of hippocampus in rats. *Toxicol Lett* 2009;184: 121-125.

Martinez L, Jimenez V, Garcia-Sepulveda C, Ceballos F, Delgado JM, Nino-Moreno P, Doniz L, Saavedra-Alanis V, Castillo CG, Santoyo ME, Gonzalez-Amaro R, Jimenez-Capdeville ME. Impact of early developmental arsenic exposure on promotor CpG-island methylation of genes involved in neuronal plasticity. *Neurochem Int* 2011; 58: 574-581.

Nisselbaum JS, Green S. A simple ultramicro method for determination of pyridine nucleotides in tissues. *Anal Biochem* 1969; 27: 212-217.

Ogata N, Ueda K, Hayaishi O. ADP-ribosylation of histone H2B. Identification of glutamic acid residue 2 as the modification site. *J Biol Chem* 1980; 255: 7610-7615.

Ogata N, Ueda K, Kagamiyama H, Hayaishi O. ADP-ribosylation of histone H1. Identification of glutamic acid residues 2, 14, and the COOH-terminal lysine residue as modification sites. *J Biol Chem* 1980; 255: 7616-7620.

Park SD, Kim CG, Kim MG. Inhibitors of poly(ADP-ribose) polymerase enhance DNA strand breaks, excision repair, and sister chromatid exchanges induced by alkylating agents. *Environ Mutagen* 1983; 5: 515-525.

Paxinos G, Watson C. *The rat brain in stereotaxic coordinates*. Academic Press, New York. 2005.

Poonepalli A, Balakrishnan L, Khaw AK, Low GK, Jayapal M, Bhattacharjee RN, Akira S, Balajee AS, Hande MP. Lack of poly(ADP-ribose) polymerase-1 gene product enhances cellular sensitivity to arsenite. *Cancer Res* 2005; 65: 10977-10983.

Qin XJ, Hudson LG, Liu W, Timmins GS, Liu KJ. Low concentration of arsenite exacerbates UVR-induced DNA strand breaks by inhibiting PARP-1 activity. *Toxicol Appl Pharmacol* 2008; 232: 41-50.

Rios R, Zarazua S, Santoyo ME, Sepulveda-Saavedra J, Romero-Diaz V, Jimenez V, Perez-Severiano F, Vidal-Cantu G, Delgado JM, Jimenez-Capdeville ME. Decreased nitric oxide markers and morphological changes in the brain of arsenic-exposed rats. *Toxicol* 2009; 261: 68-75.

Riquelme PT, Burzio LO, Koide SS. ADP ribosylation of rat liver lysine-rich histone in vitro. *J Biol Chem* 1979; 254: 3018-3028.

Rodríguez VM, Dufour L, Carrizales L, Díaz-Barriga F, Jiménez-Capdeville M.E. Effects of oral exposure to a mining waste on in vivo striatal dopamine release. *Environ Health Perspect* 1998; 106: 487-491,

Rodriguez VM, Jimenez-Capdeville ME, Giordano M. The effects of arsenic exposure on the nervous system. *Toxicol Lett* 2003; 145: 1-18.

Rosado JL, Ronquillo D, Kordas K, Rojas O, Alatorre J, Lopez P, Garcia-Vargas G, Del Carmen Caamano M, Cebrian ME, Stoltzfus RJ. Arsenic exposure and cognitive performance in Mexican schoolchildren. *Environ Health Perspect* 2007; 115: 1371-1375.

Sahgal. *Behavioural Neuroscience*. A Practical Approach, 1 and 2. Oxford University Press Inc, New York.1993.

Satchell MA, Zhang X, Kochanek PM, Dixon CE, Jenkins LW, Melick J, Szabo C, Clark RS. A dual role for poly-ADP-ribosylation in spatial memory acquisition after traumatic brain injury in mice involving NAD+ depletion and ribosylation of 14-3-3gamma. *J Neurochem* 2003; 85: 697-708.

Tseng HP, Wang YH, Wu MM, The HW, Chiou HY, Chen CJ. Association between chronic exposure to arsenic and slow nerve conduction velocity among adolescents in Taiwan. *J Health Popul Nutr* 2006; 24: 182-189.

Visochek L, Steingart RA, Vulih-Shultzman I, Klein R, Priel E, Gozes I, Cohen-Armon M. PolyADP-ribosylation is involved in neurotrophic activity. *J Neurosci* 2005; 25: 7420-7428.

Walter I, Schwerdtle T, Thuy C, Parsons JL, Dianov GL, Hartwig A. Impact of arsenite and its methylated metabolites on PARP-1 activity, PARP-1 gene expression and poly(ADP-ribosyl)ation in cultured human cells. *DNA Repair* 2007; 6: 61-70.

Wang X, Meng D, Chang Q, Pan J, Zhang Z, Chen G, Ke Z, Luo J, Shi X. Arsenic inhibits neurite outgrowth by inhibiting the LKB1-AMPK signaling pathway. *Environ Health Perspect* 2010; 118: 627-634.

Wang Y, Li S, Piao F, Hong Y, Liu P, Zhao Y. Arsenic down-regulates the expression of Camk4, an important gene related to cerebellar LTD in mice *Neurotoxicol Teratol* 2009; 31: 318-322

Wasserman GA, Liu X, Parvez F, Ahsan H, Factor-Litvak P, Kline J, van Geen A, Slavkovich V, Loiacono NJ, Levy D, Cheng Z, Graziano JH. Water arsenic exposure and intellectual function in 6-year-old children in Araihazar, Bangladesh. *Environ Health Perspect* 2007; 115: 285-289.

Zarazua S, Perez-Severiano F, Delgado JM, Martinez LM, Ortiz-Perez D, Jimenez-Capdeville ME. Decreased nitric oxide production in the rat brain after chronic arsenic exposure. *Neurochem Res* 2006; 31: 1069-1077.

Zarazua S, Rios R, Delgado JM, Santoyo ME, Ortiz-Perez D, Jimenez-Capdeville ME. Decreased arginine methylation and myelin alterations in arsenic exposed rats. *Neurotoxicol* 2010; 31: 94-100.

In: Arsenic　　　　　　　　　　　　　　　　ISBN: 978-1-63321-054-7
Editor: Marissa Jane Olson　　　　　　　© 2014 Nova Science Publishers, Inc.

Chapter 6

DISTRIBUTION AND ABUNDANCE OF ARSENIC IN THE SOILS AND PLANTS

Ismail M. M. Rahman[1,2,*] *Zinnat A. Begum*[2],
S. Y. Salehi-Lisar[3]*, Rouhollah Motafakkerazad*[3]*,
M. Rabiul Awual*[4] *and Hiroshi Hasegawa*[5]

[1]Department of Applied and Environmental Chemistry,
Faculty of Science, University of Chittagong, Chittagong, Bangladesh
[2]Graduate School of Natural Science and Technology,
Kanazawa University, Kakuma, Kanazawa, Japan
[3]Department of Plant Sciences, Faculty of Natural Sciences,
University of Tabriz, Tabriz, Iran
[4]Actinide Coordination Chemistry Group,
Quantum Beam Science Directorate,
Japan Atomic Energy Agency (SPring-8), Hyogo, Japan
[5]Institute of Science and Engineering,
Kanazawa University, Kakuma, Kanazawa, Japan

---

[*] Author for correspondence. E-mail: I.M.M.Rahman@gmail.com

ABSTRACT

Arsenic (As) has evoked concerns related to the environmental and human health issues for decades. In recent years, the concern has been relocated to the front position as more of the world's population relies on groundwater as a source of clean drinking water, which is reported to be contaminated due to the elevated level of arsenic. Arsenic, being a naturally-occurring element in the Earth's crust, has been commonly found as a trace constituent of rocks, soils, sediments, water, and biota. The natural or anthropogenic or both the activities can elevate arsenic concentrations in groundwater, soils, and sediments to toxic levels and consequently, in the plants. Arsenic exists in multiple oxidation states at earth surface conditions, while arsenite and arsenate are by far the most common As-species found in the environment. The surface processes such as precipitation, dissolution, adsorption, and desorption have been controlled by geochemical parameters, such as pH, Eh, ionic composition, and mineral type, and determine the mobility characteristics of arsenic and the abundance in any given location. The chapter will provide a critical review of the issues related to the distribution and abundance of arsenic in the soils and plants.

1. INTRODUCTION

Arsenic (As), which is a metalloid with a predominantly non-metallic character, exists in the soils, rocks, natural waters and organisms [1]. It comprises about 0.0005% of the earth's crust and is, respectively, 14th and 12th most abundant elements in seawater and human body [2]. Environment contamination with arsenic can be due to both geological and anthropogenic sources [3]. The toxicity potential and solubility of the arsenic species are varied with the corresponding chemical forms in the environment, which can be intra-transformed under the influence of several factors, such as interaction with the microbes and geochemical condition like pH and redox potential (Figure 1) [4-7]. Arsenic has no known biological role in plants, animals and humans [7, 8], while the carcinogenic [5, 9], genotoxic and mutagenic [6] effects of arsenic have been determined. The hyper and hypo- pigmentation, keratosis, hypertension, and cardiovascular diseases [2, 10] are clinical symptoms of arsenic poisoning [3]. The correlation between the skin, lung and bladder cancers and arsenic exposure are also documented [7, 11].

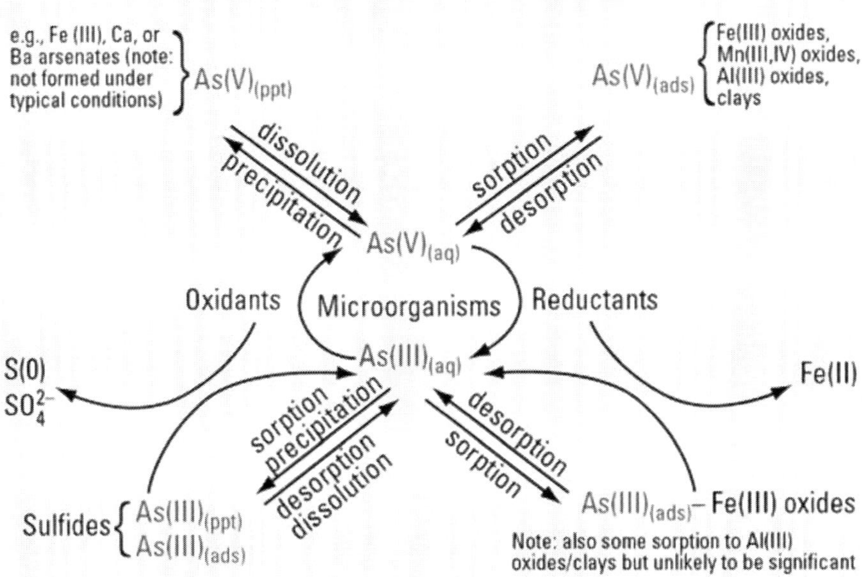

Figure 1. Possible processes in biogeochemical cycling of arsenic. "Adopted with permission from Reisinger, H. J., et al. (2005), Remediating subsurface arsenic contamination with monitored natural attenuation, *Environmental Science & Technology* 39(22): 458a-464a; ©2005, American Chemical Society."

Groundwater and food (especially marine products) are the prominent sources of arsenic ingestion in humans [4, 12, 13]. It has been estimated that millions of people are at risk of arsenic, mainly due to consumption of contaminated groundwater [2], and the most-discussed geological arsenic pollution occurrence involved the population in India and Bangladesh [14]. Arsenic contents in the foods are in the range of 0.05 to 40 μg g^{-1} [15], and the WHO provisional tolerable daily intake value of inorganic arsenic is 2.1 μg kg^{-1} body wt. day^{-1} [11]. The average intake of arsenic by diet has been estimated to be about 50 μg day^{-1} [15], while it has been varied in the range of 10 μg day^{-1} to 345 μg day^{-1} depending on the dietary habit of the populations in the different parts of the world [4].

There are several reviews about arsenic resources, contamination, distribution, behavior and toxicity in literatures, and some of those are mentioned here as the further reading resources. The review of Mandal and Suzuki [2] covers the environmental origin, occurrence, impact and toxicity of As in human and its contamination in different countries, and the global health problem due to arsenic has been reviewed by the Kinniburgh et al. [16] and Ng et al. [10]. The arsenic distribution in the environment has been discussed by Reimann et al. [17]. The source, behavior and distribution of As in natural waters has been reviewed by Smedley et al. [1], while the toxicity, speciation, and transformations of aquatic arsenic have been covered in the work of Sharma and Sohn [7]. Dembitsky and Rezanka [18] reviewed about the natural arseno-compounds in plants, while the methods and technologies of As-assay and monitoring in the environment has been compiled in the work of Melamed [5]. The focus of our current compilation will be on the abundance and the distribution aspects of arsenic in the soil and plants.

2. ARSENIC AND SOIL

The arsenic concentration is soils are generally higher than that in the rocks. The usual arsenic concentration in soils, while uncontaminated, is lowest in the sandy soils and those derived from granites, and the alluvial and organic soils have the highest [2]. Some other types of soils, such as peats and bog soils and acid sulphate soils can also contain arsenic at a higher rate [1].

The prime source of arsenic in soils is geological activity and, hence, arsenic concentrations in parent materials, e.g., minerals and rocks, are important factors [1]. Besides, climate conditions, organic and inorganic constituents in the soils are some other influencing factors on the soil arsenic content [2]. The anthropogenic arsenic input in soils comes from industrial operations, mining, burning of fossil fuel and agricultural practices (mainly pesticides and phosphate fertilizers) [1]. Distinctive concentration ranges of arsenic in soils (mg kg^{-1}) of different countries have been reported in the literatures, which indicated considerable variations among the geographic regions, such as, Argentina: 0.8–22; Bangladesh: 9.0–28; China: 0.01–626; France: 0.1–5; Germany: 2.5–4.6; Italy: 1.8–60; Japan: 0.4–70; Mexico: 2–40; South Africa: 3.2–3.7; Switzerland: 2–2.4; United States: 1.0–20; West Bengal, India: 10–196 [2, 19, 20].

Arsenic can exist in soil in both organic and inorganic species [2], while more toxic inorganic forms (arsenate and arsenite) are dominant [8, 12]. In contrast to arsenite and arsenate, some organic forms of arsenic, such as arsenosugars, are non-toxic in nature [15, 21]. Among the organic forms of arsenic that found in soils, the water-soluble forms exhibit higher toxicity [21, 22]. Therefore, arsenic in natural soils can be extremely toxic to completely non-toxic depending on the corresponding chemical forms [21].

The arsenic speciation in the natural environments, including that in the soils, is attributable to the redox potential (Eh) and acidity (pH) [7, 23]. Arsenate (As^V) is predominant in aerobic conditions [12], but under reducing conditions arsenite (As^{III}) is frequently available species [12, 24] and the ratio of As^{III} increased with the decrease of Eh [24]. Hence, the release of arsenic due to the weathering process require low pH and Eh conditions as As^{III} oxidize rapidly to insoluble As^V [17]. The As^V is stable and can easily be retained into soil solid phases, such as clays, iron and manganese oxides/hydroxides and organic materials. The Fe- and Al-arsenate, as occurred in the acidic soils, have lower solubility in comparison with some other arsenate salts, such as calcium arsenate, that occur in the alkaline and calcareous soils [2].

Besides physicochemical parameters of soils, the activity of the living organisms can alter the speciation of As. Both the As^{III} and As^V, which are the dominant As-species in the soil solution, can alter through the biological processes, e.g., oxidation, reduction [25] and methylation reactions [12]. For example, the bio-methylation of the As-species can be initiated by *Escherichia coli* [25]. Furthermore, the As^V can be reduced by the cytoplasmic arsenate reductase in bacteria, and some bacteria can sue arsenate as a terminal electron acceptor and hence reduce it. The arsenate reduction by soil bacteria can lead to releasing of more mobile arsenite to the soil solution from solid phase [24].

A low amount of As^{III} or As^V can replace other cations, such as, Si^{IV}, Al^{III}, Fe^{III} and Ti^{IV} in some minerals. For example, As^{III} can replace Fe^{III} and Al^{III} in silicates and As^V can replace P^V in minerals like apatite (a group of phosphate minerals) [17]. The arsenic and phosphate have similarities in chemical properties, and hence they compete during the absorption on surfaces of soil particles [7]. The competing trend depends on pH and initial concentrations of ions, and can lead to leaching of arsenic from contaminated soils, particularly due to the application of phosphate fertilizers [17].

3. ARSENIC AND PLANTS

Arsenic like other toxic elements can enter to plant cells via same uptake pathways of essential nutrients [22]. Arsenic has high similarity to phosphate, which is an important nutrient in plants having structural, metabolic and regulatory functions [26]. Hence, the phosphate transport system in the plants usually works during the process of As-uptake. Accordingly, arsenic corporate in some phosphorous reactions and disrupt phosphate metabolism [27], e.g., phosphorylation reactions, including ATP synthesis [18, 28]. However, the As-uptake mechanism was not similar in all studied plant species and was also different for As^{III} and As^{V}. For example, arsenite uptake took place mainly by members of aquaporins (NIP: nodulin 26-like intrinsic protein) [29], and the role of OsNIP2;1/Lsi1 (a silicon transporter) in its uptake was illustrated in rice plants [30]. Generally, the majority of the arsenic in plants is accumulated in the roots, and only a small amount is transported to the shoots [31]. Besides of inorganic forms, some organic arsenicals, e.g., arsenosugars, are important As-containing species found among the plants, such as algae [18, 24] (Figure 2).

Arsenic concentration in plants depends on several factors, which include the growth and development stage of the plant species, soil physicochemical properties, arsenic mobility in the soils, and concentration of other ions, such as phosphorous, in the soil [2, 7, 22, 26, 27]. Although arsenic exclusion is the main tolerance mechanism in most plants [28], very higher concentrations has been reported for some plant species, e.g., 6640 ppm in *Jasione montana*, 4130 ppm in *Calluna vulgaris* and 3470 ppm in *Agrostis tenuis* from mine site of England [22]. Some plant species are considered as arsenic hyperaccumulator and arsenic concentration in these plants is above 0.1% (1000 mg kg^{-1}) dry weight [18, 26] with no toxicity symptoms [28]. Until 2010, only 12 As-hyperaccumulator plant species have been identified [9]. Although accumulations of some heavy metals in some species of ferns such as mosses have been shown [32], the *Pteridaceae* family of ferns is the well-known example that includes arsenic hyperaccumulator species [9, 33]. Up to 22,000 mg kg^{-1} (2.2%) of the arsenic on a dry weight basis was reported in *Pteris vittata* [24, 26], which is a terrestrial As-hyperaccumulator plant belongs to *Pteridaceae* family [9, 26] and the order of As-content in its different organs of this plant is: leaves > leafstalks > roots [34]. An hyperaccumulator moss *Pohlia wahlenbergii* reported to contain 3% dry weight of arsenic [35]. The *Agrostis tenuis*, *Agrostis stolonifera*, *Pityrogramma calomelanos*, *Pteris longifolia*, *Pteris umbros* [24] and *Pteris cretica* [9, 24, 36] are some well-known As-hyperaccumulator plant species. Hyperaccumulator plants are good ecological indicators of arsenic contaminated environments such as mines [32] and can be used for the removal (phytoremediation) of arsenic from polluted sites [9, 26]. The hyperaccumulator plants with higher growth rate and above-ground biomass, short life cycle and higher propagation rate are good candidates for phytoremediation [24].

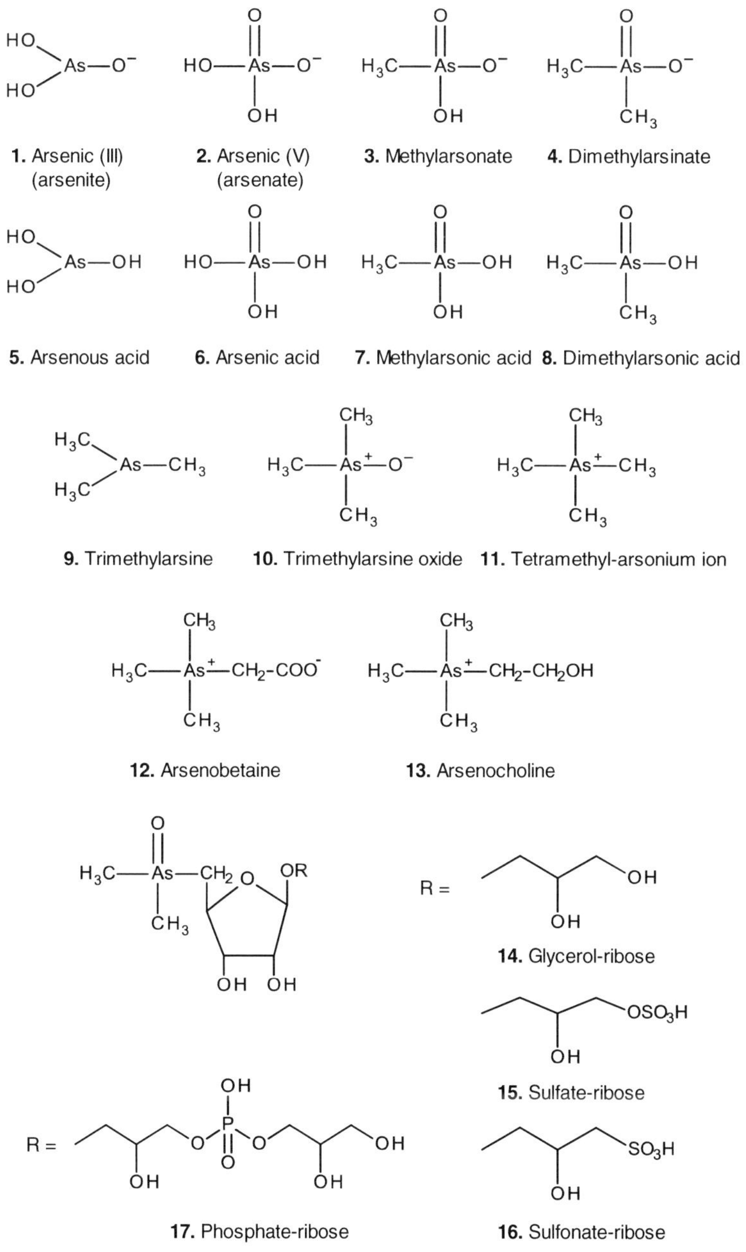

Figure 2. Inorganic and organoarsenic species isolated from plants, lichens and fungi. "Adopted with permission from Dembitsky, V. M. and T. Rezanka (2003), Natural occurrence of arseno compounds in plants, lichens, fungi, algal species, and microorganisms, *Plant Science* 165(6): 1177–1192; ©2003, Elsevier."

The hyper tolerance mechanism of arsenite was different from arsenate, e.g., *Holcus lanatus*. The non-hyperaccumulator plants mainly remove high fraction of arsenic from roots by efflux mechanisms but hyperaccumulator plants usually not use this mechanism and concentrate high arsenic in plant organs [28]. A higher capacity of arsenic translocation to the shoots is a factor of tolerance and as well as an essential mechanism in hyperaccumulator plants as observed in hyperaccumulator species *Pteris vittata* [3].

Arsenic toxicity in plants has been related to its interaction to biochemical functions of enzymes as well as other important bio molecules [9]. The comparative toxicity of arsenite is greater than that of arsenate in spite of the uptake of both the As^{III} and As^{V} in plants [22]. The higher toxicity of arsenite is related to its high affinity for sulfhydryl (thiol) group of some molecules specially proteins, such as glutathione, phytochelatins and so forth [3, 7, 18, 27, 28]. The inhibition of the activity enzymes, such as glutathione reductase, glutathione peroxidases, thioredoxin reductase and thioredoxin peroxidase are the results of arsenite-sulfur bond production [7], which might lead to the death of cells [27].

The As^{III} and As^{V} can induce oxidative stress in plants [28] as shown by higher malondialdehyde content in wheat plants [37, 38], and hence some aspects of arsenic toxicity in plants can be related to this effect. The enhancement of reactive oxygen species production due to arsenic toxicity has been observed in plants and a higher activity of antioxidant system for plant tolerance towards arsenic toxicity was reported, e.g., *A. capillus-veneris* [9] and wheat [37, 38]. A decrease in seed germination can happen due to the higher concentration of arsenic in plants, e.g., wheat. However, low levels of arsenic can stimulate seed germination and plant growth [38]. The plant height, crop yield, root growth and development can be affected negatively by the higher level of arsenic [39]. The wilting of leaves, violet coloration (increased anthocyanin), root discoloration and cell plasmolysis are common symptoms of arsenic toxicity in plants [40, 41]. Furthermore, arsenic has been diagnosed as a mitotic toxin due to its high affinity to thiol groups [28]. For example, decrease in mitotic index by arsenic has been reported in different plant species, such as *Allium sepa* and *Hordeum vulgare* [22].

Generally, prevention of metal uptake by roots, sequestration by organic compounds such as organic acids, compartmentation in certain organelles specially vacuole, binding to proteins such as phytochelatins (PCs), and ion efflux and exudation are major mechanisms of plant's tolerance to toxic elements [22]. Some of the above-mentioned mechanism has been involved in the resistance mechanism of plants towards arsenic. The stimulation of PCs production by arsenic was shown in some plant species such as *Rubia tinctorum* [42]. In *Arabidopsis thaliana*, PCs deficiency leads to higher sensitivity to arsenic [43]. Hence, synthesis of PCs can be a prime factor in basic arsenic tolerance in plants. The glutathione, which is another thiol containing tri-peptide with high affinity to arsenite, and formation of As^{III}–glutathione complexes can lead to arsenite detoxifying [3, 28].

CONCLUSION

The intrusion of arsenic due to the anthropogenic or geological releases at a higher rate may disrupt the natural balance or distribution pattern in the soil and plant environments. There have been variations in the exposure extent depending on the biosphere conditions or species types. The development of an intra-stimulated mechanism to tolerate the toxic effect of arsenic in plants has been observed. We hope that the discussion on the distribution and abundance of arsenic in the soils and plants, as covered in the current chapter, will provide a concise idea of the issue.

REFERENCES

[1]   Smedley, P. L.; Kinniburgh, D. G., A review of the source, behaviour and distribution of arsenic in natural waters. *Appl. Geochem.*, 2002, *17*, 517–568.

[2]   Mandal, B. K.; Suzuki, K. T., Arsenic round the world: A review. *Talanta*, 2002, *58*, 201–235.

[3]   Zhu, Y.-G.; Rosen, B. P., Perspectives for genetic engineering for the phytoremediation of arsenic-contaminated environments: From imagination to reality? *Curr. Opin. Biotechnol.*, 2009, *20*, 220–224.

[4]   Devesa, V.; Velez, D.; Montoro, R., Effect of thermal treatments on arsenic species contents in food. *Food Chem. Toxicol.*, 2008, *46*, 1–8.

[5]   Melamed, D., Monitoring arsenic in the environment: A review of science and technologies with the potential for field measurements. *Anal. Chim. Acta*, 2005, *532*, 1–13.

[6]   Hung, D. Q.; Nekrassova, O.; Compton, R. G., Analytical methods for inorganic arsenic in water: A review. *Talanta*, 2004, *64*, 269–277.

[7]   Sharma, V. K.; Sohn, M., Aquatic arsenic: Toxicity, speciation, transformations, and remediation. *Environ. Int.*, 2009, *35*, 743–759.

[8]   Comino, E.; Fiorucci, A.; Menegatti, S.; Marocco, C., Preliminary test of arsenic and mercury uptake by *Poa annua*. *Ecol. Eng.*, 2009, *35*, 343–350.

[9]   Singh, N.; Raj, A.; Khare, P. B.; Tripathi, R. D.; Jamil, S., Arsenic accumulation pattern in 12 Indian ferns and assessing the potential of *Adiantum capillus-veneris*, in comparison to *Pteris vittata*, as arsenic hyperaccumulator. *Bioresour. Technol.*, 2010, *101*, 8960–8968.

[10]   Ng, J. C.; Wang, J.; Shraim, A., A global health problem caused by arsenic from natural sources. *Chemosphere*, 2003, *52*, 1353–1359.

[11]   Roychowdhury, T., Impact of sedimentary arsenic through irrigated groundwater on soil, plant, crops and human continuum from Bengal delta: Special reference to raw and cooked rice. *Food Chem. Toxicol.*, 2008, *46*, 2856–2864.

[12]   Mihucz, V. G.; Tatar, E.; Virag, I.; Zang, C.; Jao, Y.; Zaray, G., Arsenic removal from rice by washing and cooking with water. *Food Chem.*, 2007, *105*, 1718–1725.

[13]   Yuan, C.; Gao, E.; He, B.; Jiang, G., Arsenic species and leaching characters in tea (*Camellia sinensis*). *Food Chem. Toxicol.*, 2007, *45*, 2381–2389.

[14]   Ahmad, K., Report highlights widespread arsenic contamination in Bangladesh. *Lancet*, 2001, *358*, 133-133.

[15]   Del Razo, L. M.; Garcia-Vargas, G. G.; Garcia-Salcedo, J.; Sanmiguel, M. F.; Rivera, M.; Hernandez, M. C.; Cebrian, M. E., Arsenic levels in cooked food and assessment of adult dietary intake of arsenic in the Region Lagunera, Mexico. *Food Chem. Toxicol.*, 2002, *40*, 1423–1431.

[16]   Gaus, I.; Kinniburgh, D. G.; Talbot, J. C.; Webster, R., Geostatistical analysis of arsenic concentration in groundwater in Bangladesh using disjunctive kriging. *Environ. Geol.*, 2003, *44*, 939–948.

[17]   Reimann, C.; Matschullat, J.; Birke, M.; Salminen, R., Arsenic distribution in the environment: The effects of scale. *Appl. Geochem.*, 2009, *24*, 1147–1167.

[18]   Dembitsky, V. M.; Rezanka, T., Natural occurrence of arseno compounds in plants, lichens, fungi, algal species, and microorganisms. *Plant Sci.*, 2003, *165*, 1177–1192.

[19]   Chakraborti, D.; Basu, G. K.; Biswas, B. K.; Chowdhury, U. K.; Rahman, M. M.; Paul, K.; Chowdhury, T. R.; Chanda, C. R.; Lodh, D., Characterization of arsenic bearing sediments in Gangetic delta of West

Bengal-India. In *Arsenic Exposure and Health Effects*, Chappell, W. R.; Abernathy, C. O.; Calderon, R. L., Eds. Elsevier: Amsterdam, 2001; pp 27–52.

[20]   Nickson, R. T.; McArthur, J. M.; Ravenscroft, P.; Burgess, W. G.; Ahmed, K. M., Mechanism of arsenic release to groundwater, Bangladesh and West Bengal. *Appl. Geochem.*, 2000, *15*, 403–413.

[21]   Soeroes, C.; Kienzl, N.; Ipolyi, I.; Dernovics, M.; Fodor, P.; Kuehnelt, D., Arsenic uptake and arsenic compounds in cultivated *Agaricus bisporus*. *Food Control*, 2005, *16*, 459–464.

[22]   Patra, M.; Bhowmik, N.; Bandopadhyay, B.; Sharma, A., Comparison of mercury, lead and arsenic with respect to genotoxic effects on plant systems and the development of genetic tolerance. *Environ. Exp. Bot.*, 2004, *52*, 199–223.

[23]   Baeyens, W.; de Brauwere, A.; Brion, N.; De Gieter, M.; Leermakers, M., Arsenic speciation in the River Zenne, Belgium. *Sci. Total Environ.*, 2007, *384*, 409–419.

[24]   Wang, S. L.; Zhao, X. Y., On the potential of biological treatment for arsenic contaminated soils and groundwater. *J. Environ. Manag.*, 2009, *90*, 2367–2376.

[25]   Stolz, J. F.; Oremland, R. S., Bacterial respiration of arsenic and selenium. *FEMS Microbiol. Rev.*, 1999, *23*, 615–627.

[26]   Robinson, B.; Kim, N.; Marchetti, M.; Moni, C.; Schroeter, L.; van den Dijssel, C.; Milne, G.; Clothier, B., Arsenic hyperaccumulation by aquatic macrophytes in the Taupo Volcanic Zone, New Zealand. *Environ. Exp. Bot.*, 2006, *58*, 206–215.

[27]   Sun, Y. H.; Li, Z. J.; Guo, B.; Chu, G. X.; Wei, C. Z.; Liang, Y. C., Arsenic mitigates cadmium toxicity in rice seedlings. *Environ. Exp. Bot.*, 2008, *64*, 264–270.

[28]   Verbruggen, N.; Hermans, C.; Schat, H., Mechanisms to cope with arsenic or cadmium excess in plants. *Curr. Opin. Plant Biol.*, 2009, *12*, 364–372.

[29]   Bienert, G. P.; Thorsen, M.; Schussler, M. D.; Nilsson, H. R.; Wagner, A.; Tamas, M. J.; Jahn, T. P., A subgroup of plant aquaporins facilitate the bi-directional diffusion of $As(OH)_3$ and $Sb(OH)_3$ across membranes. *BMC Biol.*, 2008, *6*, 26.

[30]   Ma, J. F.; Yamaji, N.; Mitani, N.; Xu, X. Y.; Su, Y. H.; McGrath, S. P.; Zhao, F. J., Transporters of arsenite in rice and their role in arsenic accumulation in rice grain. *Proc. Natl. Acad. Sci. U. S. A.*, 2008, *105*, 9931–9935.

[31] Padmavathiamma, P. K.; Li, L. Y., Phytoremediation technology: Hyper-accumulation metals in plants. *Water Air Soil Pollut.*, 2007, *184*, 105–126.

[32] Chang, J.-S.; Yoon, I.-H.; Kim, K.-W., Heavy metal and arsenic accumulating fern species as potential ecological indicators in As-contaminated abandoned mines. *Ecol. Indic.*, 2009, *9*, 1275–1279.

[33] Gumaelius, L.; Lahner, B.; Salt, D. E.; Banks, J. A., Arsenic hyperaccumulation in gametophytes of *Pteris vittata*. A new model system for analysis of arsenic hyperaccumulation. *Plant Physiol.*, 2004, *136*, 3198–3208.

[34] Chen, T. B.; Wei, C. Y.; Huang, Z. C.; Huang, Q. F.; Lu, Q. G.; Fan, Z. L., Arsenic hyperaccumulator *Pteris vittata* L. and its arsenic accumulation. *Chin. Sci. Bull.*, 2002, *47*, 902–905.

[35] Craw, D.; Rufaut, C.; Haffert, L.; Paterson, L., Plant colonization and arsenic uptake on high arsenic mine wastes, New Zealand. *Water Air Soil Pollut.*, 2007, *179*, 351–364.

[36] Fayiga, A. O.; Ma, L. Q., Arsenic uptake by two hyperaccumulator ferns from four arsenic contaminated soils. *Water Air Soil Pollut.*, 2005, *168*, 71–89.

[37] Liu, X.; Zhang, S.; Shan, X.-q.; Christie, P., Combined toxicity of cadmium and arsenate to wheat seedlings and plant uptake and antioxidative enzyme responses to cadmium and arsenate co-contamination. *Ecotoxicol. Environ. Saf.*, 2007, *68*, 305–313.

[38] Li, C.-x.; Feng, S.-l.; Shao, Y.; Jiang, L.-n.; Lu, X.-y.; Hou, X.-l., Effects of arsenic on seed germination and physiological activities of wheat seedlings. *J. Environ. Sci.*, 2007, *19*, 725–732.

[39] Arain, M. B.; Kazi, T. G.; Baig, J. A.; Jamali, M. K.; Afridi, H. I.; Shah, A. Q.; Jalbani, N.; Sarfraz, R. A., Determination of arsenic levels in lake water, sediment, and foodstuff from selected area of Sindh, Pakistan: Estimation of daily dietary intake. *Food Chem. Toxicol.*, 2009, *47*, 242–248.

[40] Hossain, M. F., Arsenic contamination in Bangladesh - An overview. *Agric. Ecosyst. Environ.*, 2006, *113*, 1–16.

[41] Huq, S. M. I.; Shila, U. K.; Joardar, J. C., Arsenic mitigation strategy for rice, using water regime management. *Land Contam. Reclaim.*, 2006, *14*, 805–813.

[42] Maitani, T.; Kubota, H.; Sato, K.; Yamada, T., The composition of metals bound to class III metallothionein (phytochelatin and its

desglycyl peptide) induced by various metals in root cultures of *Rubia tinctorum*. *Plant Physiol.*, 1996, *110*, 1145–1150.

[43] Ha, S.-B.; Smith, A. P.; Howden, R.; Dietrich, W. M.; Bugg, S.; O'Connell, M. J.; Goldsbrough, P. B.; Cobbett, C. S., Phytochelatin synthase genes from arabidopsis and the yeast *Schizosaccharomyces pombe*. *Plant Cell*, 1999, *11*, 1153–1163.

INDEX

D

E

F

G

H

I

J

K

L

S